Lecture Notes in Mathematics 1492

Lecture Notes in Mathematics

Editors:
A. Dold, Heidelberg
B. Eckmann, Zürich
F. Takens, Groningen

Keith R. Wicks

Fractals and Hyperspaces

Springer-Verlag
Berlin Heidelberg New York
London Paris Tokyo
Hong Kong Barcelona
Budapest

Author

Keith R. Wicks
Department of Mathematics and Computer Science
University College of Swansea
Singleton Park
Swansea SA2 8PP, U. K.

The picture on the front cover shows a zoom-in on Fig. 14, page 51

Mathematics Subject Classification (1991): 03H05, 05B45, 51N05, 52A20, 52A45, 54A05, 54B20, 54C60, 54E35, 54E40, 54H05, 54H20, 54H25, 54J05

ISBN 3-540-54965-X Springer-Verlag Berlin Heidelberg New York
ISBN 0-387-54965-X Springer-Verlag New York Berlin Heidelberg

Typesetting: Camera ready by author
Printing and binding: Druckhaus Beltz, Hemsbach/Bergstr.
46/3140-543210 - Printed on acid-free paper

Foreword

The main theme of this monograph is the study of fractals and fractal notions, backed up by a self-contained nonstandard development of relevant hyperspace theory, particularly as regards the Hausdorff metric and Vietoris topology. The fractal study itself is in two parts, the first developing and making contributions to the theory of J. E. Hutchinson's invariant sets, sets which are self-similar in the sense of being composed of smaller images of themselves. The second part explores newer territory, introducing the formal notion of a 'view' as part of a general framework concerned with studying the structure and perception of sets within a given space, and in particular we use views to express and investigate new concepts of self-similarity and fractality which are then considered in connection with invariant sets, a large class of which are shown to be 'visually fractal' in a certain precise sense. Complete with many figures and suggestions for further work, the monograph should be of relevance to those interested in fractals, hyperspaces, fixed-point theory, tilings, or nonstandard analysis.

The work was undertaken at the University of Hull during the period 1987-90, financed for two years by an SERC research grant which I gratefully acknowledge. My grateful thanks go also to Professor Nigel Cutland for assistance and advice throughout, and to Dr. Dona Strauss for help with many topological and other queries.

Keith R. Wicks
Hull, August 1991.

Foreword

Contents

Chapter 4 : Views and Fractal Notions

Appendices

Introduction

As briefly outlined in the Foreword, the main theme of this monograph is the study of fractals and fractal notions, namely comprising Chapters 3 and 4, aided by the prior development of certain hyperspace theory in Chapters 1 and 2. To understand the motivation for the latter, consider now the first part of the fractal study.

The notion of self-similarity has always been one of the central themes in the subject of fractals, and in Chapter 3 we consider the particular type of self-similarity captured by J. E. Hutchinson's 'invariant sets'. These are nonempty compact subsets K of a complete metric space X, which are composed of smaller images of themselves in the sense that $K = \bigcup_{f \in F} fK$ for some finite set F of contractions of X. Strictly speaking it is not so much K that is studied, rather the properties of K dictated by F. A general treatment is given, blending in the already existing basic results with contributions of the author, one aspect of which is the generalization from *finite* to certain *compact* sets F of contractions (or even more generally, 'reductions'). Further aspects include consideration of the regularity or residuality of invariant sets, and a study of those which are *tiled* by their images under the maps involved.

The natural setting for invariant sets is the space $\mathscr{K}X$ of nonempty compact subsets of X equipped with the Hausdorff metric; indeed, the *existence* of invariant sets follows from an elegant application in $\mathscr{K}X$ of Banach's contraction mapping theorem. We therefore precede the study of invariant sets with a chapter introducing the Hausdorff metric and using nonstandard analysis to quickly prove everything needed later on. In particular we give short nonstandard proofs of the facts that if X is complete so is $\mathscr{K}X$, and that if X is boundedly compact so is $\mathscr{K}X$ (the latter result being the Blaschke Selection Theorem in the case $X = \mathbb{R}^n$). In turn however, since the *topology* of the Hausdorff metric is an example of the 'Vietoris topology' defined in terms of the *topology* of X, the study of the topological aspect of $\mathscr{K}X$ is relegated to a preceding chapter on the Vietoris topology, which is defined on the nonempty compact subsets (or more generally, on the *closed* subsets) of *any* Hausdorff space. Extensive use is again made of nonstandard analysis, demonstrating its potential in the study of hyperspaces and intended to be as much a part of the work as the results themselves, many of which are in any case well-known, such as the results that if X is compact or locally compact, so respectively is $\mathscr{K}X$.

Hyperspace theory is also drawn upon in the fourth and final chapter which, comprising the second part of the fractal work, introduces the formal concept of a 'view' as part of a general framework concerned with studying the structure and perception of sets within a given space X. Roughly speaking, a 'view' of a subset A of X consists of a region of X together with the part of A lying *within* that region. In the case of $X = \mathbb{R}^2$ this closely models what one might expect to see upon looking down and surveying A. Using views we define a new notion of self-similarity, being a formal interpretation of the idea that wherever one looks at the set in question one can see the same sort of structure. This property is shown to hold for certain 'ω-extensions' of invariant sets, obtained roughly speaking by 'growing out' an invariant set K indefinitely. With the aid of a natural topology on a set of views we go on to define a second fractal notion, that of a subset A of $\mathbb{R}^n$ being 'visually fractal' at a point $x \in A$. This is an interpretation of the idea that as we 'zoom in' on x we see detail beyond detail; more precisely it expresses that no visual convergence takes place, and in particular we show that many residual invariant sets are visually fractal at all points. A number of other view-related notions are more briefly touched upon and may begin to make clear the wider potential application of views, facilitating what might be thought of in general as 'visual analysis'. Chapters 3 and 4 each conclude with a number of notes, questions, and suggestions for further work.

Several appendices to the monograph are provided and cover an assortment of background topics including further hyperspace work, of which we take advantage in an appendix on tilings when we give a nonstandard proof of a fairly general tiling existence theorem.

Since nonstandard analysis is used throughout in studying any topological or metric space, we provide for convenience a 'Preliminaries' section which gives a round-up of the nonstandard formulations used, including proofs of some new ones. Also included are one or two reminders of *standard* definitions and theory.

Numbering Convention : Each chapter consists of sections numbered 0, 1, 2 etc., Section 0 being a short introduction and summary of the work covered. In Section n the results are numbered n .1, n .2, n .3 etc. with the chapter number suppressed, and reference to say '2.4' will naturally mean result 2.4 of the present chapter. Occasional reference to results from *other* chapters will explicitly mention the chapter involved. After the four chapters there follow nine appendices, A1 to A9, the results in appendix n being numbered An .1, An .2 and so on.

Preliminaries

0. Introduction

In this monograph nonstandard analysis will be routinely used in the study of any topological or metric space, and the reader who wishes to follow all the proofs will therefore need to have a degree of familiarity with basic nonstandard methods. We shall be using the 'superstructure' approach, the objects O under investigation always being assumed to lie in the 'standard universe', thus having a corresponding image O^* in the '*nonstandard* universe' (O^* just being O if O is an 'atom'). The most extensive introductory accounts of this approach lie in Chapter 2 of Hurd and Loeb's book [HL], Lindstrøm's 'An Invitation to Nonstandard Analysis' [Li,3], the section on Logic and Superstructures in Keisler's book [Ke], and Davis' book [Da]. A briefer account can be found in Cutland's survey paper [Cu] however which gives a clear descripton of the *axiomatic* approach which is perhaps the *simplest* path into the nonstandard world. Another relatively brief account lies in [AFHL].

In using nonstandard methods we'll assume we always have sufficient 'saturation' for the job in hand; technically, we can just assume 'polysaturation' (see [Li,3]). In particular this implies the useful fact that every set A in the standard universe has a hyperfinite intermediary (Terminology: an **intermediary** of A is an internal set I such that $A^\sigma \subseteq I \subseteq A^*$, where A^σ denotes $\{a^* \mid a \in A\}$. Such a set is thus *intermediate* between the pointwise image A^σ of A and the $*$-transform A^* of A; and being *internal* we have a handle on it. As yet the literature seems to have gone without a term for the concept and by now its introduction is well overdue). A further consequence is that for any lower-directed set $(X,\leq)$ in the standard universe (lower-directed meaning that $\leq$ is a partial ordering of nonempty X with respect to which every two elements of X have a common lower bound), every intermediary I of X contains a lower bound of X^σ. In the following few pages we'll give a round-up of most of the nonstandard formulations of various topological and metric concepts used in the text, although a few will be left until required. Many can be found in one or more of the nonstandard sources listed above, along with [SL] and the original book [Ro] of Robinson, but there are one or two which as far as I know have not appeared in print; in particular, formulations of when a function between spaces is open, closed, or perfect. The round-up will also serve to fix our notation and to give one or two reminders of *standard* definitions and theory. Two central nonstandard definitions and results to have in mind are the following. Suppose X is a set in the standard universe and $\mathcal{A}$ is a nonempty collection of subsets of X. If $\mathcal{A}$ is a filter base (i.e. is lower-directed w.r.t. $\subseteq$), the **intersection monad** of $\mathcal{A}$ is $\mu\mathcal{A} = \bigcap\{A^* \mid A \in \mathcal{A}\}$, and saturation gives that for any internal $B \subseteq X^*$, $\mu\mathcal{A} \subseteq B \Leftrightarrow \exists A \in \mathcal{A}\;\; A^* \subseteq B$. On the other hand, if $\mathcal{A}$ is an ideal base (i.e. is upper-directed w.r.t $\subseteq$), the **union monad** of $\mathcal{A}$ is $\nu\mathcal{A} = \bigcup\{A^* \mid A \in \mathcal{A}\}$, and for B as above, $B \subseteq \nu\mathcal{A} \Leftrightarrow \exists A \in \mathcal{A}\;\; B \subseteq A^*$.

Unless otherwise stated, assume below that X is a topological space, and that $x \in X$, $A \subseteq X$, $\alpha \in X^*$ and $B \subseteq X^*$. Throughout the monograph, $\mathscr{C}X$ will denote the set of closed subsets of X, and $\mathscr{K}X$ the set of nonempty compact subsets of X.

1. Monads

Where $\mathscr{N}x$ is the neighbourhood filter of x, the **monad** of x is $\mu x = \bigcap\{N^* \mid N \in \mathscr{N}x\}$, the intersection monad of $\mathscr{N}x$, equivalently the intersection monad of the set $\mathscr{O}x$ of *open* neighbourhoods of x. We say α is **near** x (more strictly, near x^*) if $\alpha \in \mu x$. If α is near some x, α is **nearstandard** (the 'standard' points of X^* being those of the form x^* for $x \in X$), otherwise it is **remote**. The set of nearstandard points of X^* is denoted by $ns\, X^*$, the remote points by $rm\, X^*$. For *internal* B, by saturation

$$(\exists N \in \mathscr{N}x \quad N^* \subseteq B) \Leftrightarrow \mu x \subseteq B\text{, and}$$

$$(\forall N \in \mathscr{N}x \quad B \text{ intersects } N^*) \Leftrightarrow B \text{ intersects } \mu x\,.$$

Putting $B = A^*$ gives the nonstandard formulation of interior and limit points ;

$$x \in \underline{A} \Leftrightarrow \mu x \subseteq A^*\text{, and}$$

$$x \in \overline{A} \Leftrightarrow A^* \text{ intersects } \mu x\,.$$

The monads thus *completely determine* the topology of X, and to the nonstandard analyst form its prime description. Another result of saturation is that $\exists N \in (\mathscr{N}x)^* \ N \subseteq \mu x$. Points x and y are separated (i.e. have disjoint neighbourhoods) iff μx and μy are disjoint. Thus X is Hausdorff iff $x \neq y \Rightarrow \mu x$ and μy are disjoint. In this case, any nearstandard α is near a *unique* x, and this x is called the **standard part** of α, denoted by $^{\circ}\alpha$. In the literature $^{\circ}\alpha$ is frequently written also as '$st\ \alpha$', but we shall avoid this for reasons described below.

2. Standard and Substandard Parts of Sets

The **standard part** of B is $st\, B = \{x \in X \mid \exists b \in B \quad b \in \mu x\}$ whilst the **substandard part** of B is $subst\, B = \{x \in X \mid \mu x \subseteq B\} \subseteq st\, B$. Unlike standard parts of sets, *substandard* parts do not seem to have made it into the literature as yet, but they are *dual* to the notion of standard parts in the sense that if B_1 and B_2 are complements in X^* then $st\, B_1$ and $subst\, B_2$ are complements in X. For *internal* B, $st\, B$ is closed and dually $subst\, B$ is open. Note that $\overline{A} = st\,(A^*)$ and $\underline{A} = subst\,(A^*)$. The **monadic cover** of A is $A^{\mu} = \bigcup_{a \in A} \mu a$ (another definition not in the literature). Note that $X^{\mu} = ns\, X^*$.

We shall often be dealing with spaces Y ('hyperspaces') whose elements are subsets of another space X, and hence an element $B \in Y^*$ may have a standard part $^{\circ}B$ in Y; but this need *not* be $st\, B$ (defined since $B \subseteq X^*$), which has a quite separate definition. This explains why we avoid having 'st' used synonymously for $^{\circ}$. And to clear up any *verbal* ambiguity one can read $st\, B$ as the *pointwise* standard part of B.

3. Monads of Subsets

The idea of the monad of a *point* in X can be generalized to the monad of a *subset* of X. Where $\mathcal{N}A$ is the neighbourhood filter of A ($N \subseteq X$ being a neighbourhood of A iff $A \subseteq \underline{N}$), the **monad** of A is $\mu_X A = \bigcap\{N^* \mid N \in \mathcal{N}A\}$, the intersection monad of $\mathcal{N}A$. We use $\mu_X A$ rather than μA due again to our future involvement with hyperspaces; μA will always be reserved to denote the monad of A as an element of a hyperspace, and is distinct from $\mu_X A$ (which is a subset of X^*, unlike μA which would be a *set* of subsets of X^*). Note that $\mu x = \mu_X\{x\}$. Also, $A_1 \subseteq A_2 \Rightarrow \mu_X A_1 \subseteq \mu_X A_2$, and in general A^* and A^μ are subsets of $\mu_X A$ ($\mu_X A$ equalling A^* iff A is open). A_1 and A_2 are separated iff $\mu_X A_1$ and $\mu_X A_2$ are disjoint. In particular, x and A are separated iff μx and $\mu_X A$ are disjoint, i.e. $x \notin st\, \mu_X A$. From this it follows that X is regular iff $\forall A \in \mathcal{C}X \;\; st\, \mu_X A = A$ (using that we *always* have $A \subseteq st\, \mu_X A$ for closed A, as $A = st\, A^* \subseteq st\, \mu_X A$); equivalently, $\forall A \subseteq X \;\; st\, \mu_X A = \overline{A}$.

4. Subspaces

Any subset A of X is itself a topological space, a subspace of X, and the monads with respect to it have the following simple relation to those with respect to X. For $a \in A$, the monad of a with respect to the space A is easily found to be $A^* \cap \mu a$ (i.e. an element $\alpha \in A^*$ is near a in A^* iff it's near a in X^*), which we'll denote by $\mu_A a$.

5. Compactness and Local Compactness

X is compact iff every element of X^* is nearstandard. Using the formulation of monads of a subspace A, compactness of A can be expressed directly in terms of its monads in X by: A is compact iff $A^* \subseteq A^\mu$ (i.e. every element of A^* is near an element of A). Frequently useful is the alternative formulation that A is compact iff $\mu_X A = A^\mu$ (recall that in general we can only say $A^\mu \subseteq \mu_X A$). If X is regular, $st\, B$ is compact for any internal $B \subseteq ns\, X^*$. ***Cp X*** denotes the set of compact subsets of X, and the set of **compact** points of X^* is $\boldsymbol{cp}\, X^* = \bigcup\{K^* \mid K \in Cp\, X\}$, the union monad of $Cp\, X$. We'll say $A \subseteq X$ is **subcompact** if it's a subset of a compact set. The set ***subCp X*** of subcompact sets is thus the ideal generated by the compact sets, so that $cp\, X^*$ is also the union monad of the subcompact sets. Note that $cp\, X^* \subseteq ns\, X^*$ (since for compact K, $K^* \subseteq K^\mu \subseteq ns\, X^*$). A point x has a compact neighbourhood iff $\mu x \subseteq cp\, X^*$. Hence X is locally compact iff $cp\, X^* = ns\, X^*$, i.e. every nearstandard point is compact.

6. Convergence of Sequences

Consider a sequence $x = \langle x_n \rangle$ in X. Formally x is a function $x : \omega \to X$, so $x^* : \omega^* \to X^*$. For $n \in \omega^*$ we'll usually write x^*_n (the value of x^* at n) as x_n in line with the normal convention regarding $*$-transforms of functions f, of abbreviating $f^*\alpha$ to $f\alpha$. Then $\langle x_n \rangle \to y \Leftrightarrow \forall$ infinite $n \in \omega^*$ $x_n \in \mu y$.

7. Properties of Functions

Let $f: X \to Y$, Y being another topological space. Then f is continuous at x iff $\forall \alpha \in \mu x$ $f\alpha \in \mu fx$, i.e. $f\mu x \subseteq \mu fx$. Thus f is continuous iff $\forall x$ $f\mu x \subseteq \mu fx$. From this it follows that for bijective f, f is a homeomorphism iff $\forall x$ $f\mu x = \mu fx$, whilst for injective f, f is an embedding iff $\forall x \in X$ $\forall \alpha \in X^*$, $f\alpha \in \mu fx \Leftrightarrow \alpha \in \mu x$. On the other hand (compare with the formulation of continuity),

$$f \text{ is open } \Leftrightarrow \forall x \quad f\mu x \supseteq \mu fx.$$

(*Proof*: Suppose f is open, and take $\beta \in \mu fx$. Letting $I = \{V \in (\mathcal{O}x)^* \mid \beta \in fV\}$, I is an intermediary of $\mathcal{O}x$ (using that $\forall U \in \mathcal{O}x$ $fU \in \mathcal{O}fx$), so as $(\mathcal{O}x, \subseteq)$ is lower-directed, a consequence of saturation noted earlier says that I contains a lower bound V of $(\mathcal{O}x)^\sigma$, so $V \subseteq \bigcap (\mathcal{O}x)^\sigma = \mu x$, thus $\beta \in fV \subseteq f\mu x$, so $\beta \in f\mu x$ as required. Conversely, suppose $\forall x$ $f\mu x \supseteq \mu fx$, and take any open $U \subseteq X$. $\forall x \in U$, since $\mu x \subseteq U^*$ we have $\mu fx \subseteq f\mu x \subseteq f(U^*) = (fU)^*$ so $fx \in \underline{fU}$. Thus fU is open.)

The formulation of when f is *closed* suggests itself when the property is first of all given the *standard* reformulation that f 'protects fibres' (a *fibre* of f being a set of the form $f^{-1}y$) in the following sense . We say a point or set is *isolated from A* if the point or set has a neighbourhood disjoint from A (equivalently if its monad is disjoint from A^*). Then

$$\begin{aligned} f \text{ is closed} &\Leftrightarrow \forall A \subseteq X \quad \overline{fA} \subseteq f\overline{A} \\ &\Leftrightarrow \forall A \subseteq X\ \forall y \in Y\ (y \notin f\overline{A} \Rightarrow y \notin \overline{fA}) \\ &\Leftrightarrow \forall y \in Y\ \forall A \subseteq X\ (f^{-1}y \text{ disjoint from } \overline{A} \Rightarrow y \notin \overline{fA}) \\ &\Leftrightarrow \forall y \in Y\ \forall A \subseteq X\ (f^{-1}y \text{ isolated from } A \Rightarrow y \text{ isolated from } fA). \end{aligned}$$

This is perhaps the most instructive standard formulation of what 'f is closed' really means; it's really to do with the way f 'collapses' X, and it leads to the nonstandard formulation that any point $\alpha \in X^*$ mapping near $y \in Y$ is near $f^{-1}y$;

$$f \text{ is closed } \Leftrightarrow \forall y \in Y \quad f^{-1}\mu y \subseteq \mu_X f^{-1}y.$$

(*Proof*: Assume the left. Then for $\alpha \notin \mu_X f^{-1}y$ there's $N \in \mathcal{N}f^{-1}y$ with $\alpha \notin N^*$, so $\alpha \in \neg(N^*) = (\neg N)^*$ ($\neg$ denoting complement) which is disjoint from $\mu_X f^{-1}y$ (i.e. $f^{-1}y$ is isolated from $\neg N$), so $f\alpha \in f\neg N^*$ which is disjoint from μy (as y is isolated from

$f\neg N$) so $\alpha \notin f^{-1}\mu y$. Conversely, assume the right, and suppose $y \in Y$ and $A \subseteq X$ with $f^{-1}y$ isolated from A, i.e. $\mu_X f^{-1}y$ disjoint from A^*. Then $\forall \alpha \in A^*$, $\alpha \notin \mu_X f^{-1}y$ so $\alpha \notin f^{-1}\mu y$, i.e. $f\alpha \notin \mu y$; so fA^* is disjoint from μy, i.e. y is isolated from fA.)

It can be shown that f is continuous iff $\forall y \in Y\ f^{-1}\mu y \supseteq \mu_X f^{-1}y$. As a corollary to this and the last result, note that

$$f \text{ is closed and continuous} \Leftrightarrow \forall y \in Y\ \ f^{-1}\mu y = \mu_X f^{-1}y.$$

f is said to be *perfect* (also known as *proper*) (see [Mu] or [Du]) if it is closed and continuous with compact fibres $f^{-1}y$. The nonstandard formulation is that $\alpha \in X^*$ maps near $y \in Y$ iff α is near some $x \in X$ mapping to y;

$$f \text{ is perfect} \Leftrightarrow \forall y \in Y\ \ f^{-1}\mu y = (f^{-1}y)^{\mu}.$$

(*Proof*: Assuming the left, then $\forall y \in Y\ f^{-1}\mu y = \mu_X f^{-1}y = (f^{-1}y)^{\mu}$. Assuming the right, then f is closed (using $(f^{-1}y)^{\mu} \subseteq \mu_X f^{-1}y$) and continuous (which by the usual formulation amounts to asking that $\forall y \in Y\ \ f^{-1}\mu y \supseteq (f^{-1}y)^{\mu}$), and we now have $\forall y \in Y\ \mu_X f^{-1}y = f^{-1}\mu y = (f^{-1}y)^{\mu}$, so each fibre $f^{-1}y$ is compact.)

Suppose f is perfect. It easily follows from the above formulation that $\forall C \subseteq Y$ $f^{-1}(C^{\mu}) = (f^{-1}C)^{\mu}$, so f^{-1} preserves compactness of sets (for if $C \subseteq Y$ is compact, then $(f^{-1}C)^* = f^{-1}(C^*) \subseteq f^{-1}(C^{\mu}) = (f^{-1}C)^{\mu}$, so $f^{-1}C$ is compact), giving $f^{-1}cp\ Y^* \subseteq cp\ X^*$. As f preserves compactness too, we also have $f\,cp\ X^* \subseteq cp\ Y^*$, i.e. $cp\ X^* \subseteq f^{-1}cp\ Y^*$, so $f^{-1}cp\ Y^* = cp\ X^*$. In addition, $f^{-1}\,ns\ Y^* = ns\ X^*$ ($f^{-1}(Y^{\mu}) = (f^{-1}Y)^{\mu} = X^{\mu}$), and we can give the following proof of the standard result that if f is also surjective, Y is locally compact iff X is locally compact ;

$$\begin{aligned} Y \text{ is locally compact} &\Leftrightarrow ns\ Y^* = cp\ Y^* \\ &\Leftrightarrow f^{-1}\,ns\ Y^* = f^{-1}cp\ Y^* \\ &\Leftrightarrow ns\ X^* = cp\ X^* \\ &\Leftrightarrow X \text{ is locally compact.} \end{aligned}$$

8. Open Identification Maps

Recall that $f: X \to Y$ is an *identification map* if it's a surjection such that $\forall B \subseteq Y$, B is open iff $f^{-1}B$ is open. For such a map f, letting $\sim$ be the equivalence relation partitioning X into the fibres $f^{-1}y$ of f, so $x_1 \sim x_2 \Leftrightarrow fx_1 = fx_2$, the following are equivalent ;

(1) f is open.

(2) $\forall x_1, x_2 \in X\ (x_1 \sim x_2 \Rightarrow \forall \alpha_1 \in \mu x_1\ \exists \alpha_2 \in \mu x_2\ \ \alpha_1 \sim \alpha_2)$.

(3) $\forall x \in X\ \forall \alpha \in X^*\ \ (f\alpha \in \mu fx \Leftrightarrow \exists \beta \in \mu x\ \ \alpha \sim \beta)$.

($Proof$: (1) $\Rightarrow$ (3) : $f\mu x = \mu f x$, so $f\alpha \in \mu f x \Leftrightarrow f\alpha \in f\mu x \Leftrightarrow \exists\beta \in \mu x\ f\alpha = f\beta$.
(3) $\Rightarrow$ (2) : If $x_1 \sim x_2$ and $\alpha_1 \in \mu x_1$, $f\alpha_1 \in \mu f x_1 = \mu f x_2$ so $\exists \alpha_2 \in \mu x_2\ \alpha_1 \sim \alpha_2$.
(2) $\Rightarrow$ (1) : Let $V \subseteq X$ be open. We show $f V$ is open, i.e. $f^{-1} f V$ is open. Let $x \in f^{-1} f V$ and $\alpha_1 \in \mu x_1$; we must show $\alpha_1 \in f^{-1} f V^*$. There's $x_2 \in V$ with $x_1 \sim x_2$, and in turn there's $\alpha_2 \in \mu x_2 \subseteq V^*$ with $\alpha_1 \sim \alpha_2$, so $\alpha_1 \in f^{-1} f \alpha_2 \subseteq f^{-1} f V^*$ as required.)

In the case where $\sim$ is the relation induced by the action of a group G on X such that the action of each $g \in G$ is a homeomorphism, and $Y = X/\sim$, the quotient map $f: X \to Y$ is open, since if $x_1 \sim x_2$ then where $g \in G$ with $g x_1 = x_2$, we have $g \mu x_1 = \mu x_2$ so $\forall \alpha_1 \in \mu x_1\ \alpha_1 \sim g\alpha_1 \in \mu x_2$.

Finally we note that for an open identification map $f: X \to Y$,

$$Y \text{ is Hausdorff} \Leftrightarrow \sim \text{ is closed in } X \times X.$$

($Proof$: Assume the left and suppose $\alpha_1 \in \mu x_1$ and $\alpha_2 \in \mu x_2$ with $\alpha_1 \sim \alpha_2$. Then $\mu f x_1$ and $\mu f x_2$ intersect in $f\alpha_1 = f\alpha_2$ so $f x_1 = f x_2$, i.e. $x_1 \sim x_2$. Conversely assume the right, and suppose $\mu f x_1$ and $\mu f x_2$ intersect, say at $f\alpha$. Then there are $\alpha_i \in \mu x_i$ with $\alpha \sim \alpha_i$, and we then have $\alpha_1 \sim \alpha_2$, hence $x_1 \sim x_2$, i.e. $f x_1 = f x_2$.)

9. Ordering of Topologies

For topologies $\mathcal{O}_1$ and $\mathcal{O}_2$ on a set X, we say $\mathcal{O}_1$ is a *refinement* of $\mathcal{O}_2$ (written $\mathcal{O}_1 \leq \mathcal{O}_2$) or $\mathcal{O}_2$ is a *coarsening* of $\mathcal{O}_1$, if $\mathcal{O}_1 \supseteq \mathcal{O}_2$; equivalently if $\forall x \in X\ \mu_1 x \subseteq \mu_2 x$ (where $\mu_i x$ denotes the monad of x with respect to $\mathcal{O}_i$). The finer the topology, the smaller the monads. For a family of topologies $\mathcal{O}_i$ on X, the monad of x with respect to the *conjunction* (i.e. coarsest common refinement, i.e. g.l.b. with respect to $\leq$) of these topologies is the intersection of the monads $\mu_i x$.

10. Products

The monads of a product space $X = \prod_{i \in I} Y_i$ are given by $\alpha \in \mu x \Leftrightarrow \forall i \in I\ \alpha_i \in \mu x_i$.

11. Induced Topologies

Suppose X is a set, Y is a space, and $f: X \to Y$. The coarsest topology on X such that f is continuous will be called the <u>topology on X induced by f</u> and its monads are given by $\alpha \in \mu x \Leftrightarrow f\alpha \in \mu f x$.

12. Generating a Topology

For a set X and $\mathcal{V} \subseteq \mathcal{P}X$, the coarsest topology on X expanding $\mathcal{V}$ is called the <u>topology generated by $\mathcal{V}$</u>, and the monads are given by $\mu x = \bigcap\{V^* \mid V \in \mathcal{V} \text{ and } x \in V\}$, in other words $\alpha \in \mu x \Leftrightarrow \forall V \in \mathcal{V}(x \in V \Rightarrow \alpha \in V^*)$. This latter formulation will be used when, having defined a topology in terms of a generating set $\mathscr{V}$, we go on to give a formulation of the monads. Typically, the nonstandard analyst tries to choose suitable $\mathscr{V}$ in order to obtain *desired* monads.

13. Bounded Compactness of Metric Spaces

Let X be a metric space with metric d, and recall that the relation $\approx$ in X^* is defined by $\alpha \approx \beta \Leftrightarrow d(\alpha, \beta)$ is infinitesimal. We denote the set of points infinitesimally distant from α by $\alpha^{\approx}$. For $x \in X$, $\mu x = (x^*)^{\approx}$ of course. We say $\alpha \in X^*$ is bounded if it is *finitely* distant from some standard point x^*. Equivalently, where $\boldsymbol{Bd}\,X$ is the set of bounded subsets of X, the set $\boldsymbol{bd}\,X^*$ of bounded points of X^* is the union monad $\nu\, Bd\,X$ of $Bd\,X$, namely $bd\,X^* = \bigcup\{A^* \mid A \in Bd\,X\}$. A is bounded iff $A^* \subseteq bd\,X^*$. We'll say X is boundedly compact if, every closed bounded set is compact, equivalently if every bounded set is subcompact (i.e. $Bd\,X$ is the ideal generated by $Cp\,X$). The nonstandard formulation is as follows (compare with that of compactness) ;

$$X \text{ is boundedly compact} \Leftrightarrow \text{Every } \textit{bounded} \text{ element of } X^* \text{ is nearstandard.}$$

(*Proof*: Assuming the left, then every bounded point is *compact*, hence nearstandard. Assuming the right, then for any closed bounded $A \subseteq X$, $A^* \subseteq bd\,X^* \subseteq ns\,X^*$ so (as X is regular) $st\,A^*$ is compact, i.e. A is compact). In general $ns\,X^* \subseteq bd\,X^*$ of course, so the above says X is boundedly compact iff equality holds. A further *standard* formulation is that X is boundedly compact iff every bounded infinite subset of X has a cluster point. Bounded compactness implies local compactness.

14. The Compact-Open Topology

We denote the set of continuous maps $X \to Y$ by $\boldsymbol{C(X,Y)}$. For Hausdorff X and Y with X locally compact, the compact-open topology on $C(X,Y)$ is the coarsest topology on $C(X,Y)$ such that the natural evaluation map $C(X,Y) \times X \to Y$ (in which $(f,x) \mapsto fx$) is continuous. The topology is Hausdorff, with monads given by $\mu f = \{g \in C(X,Y)^* \mid \forall x \in X\ \forall \alpha \in \mu x \quad g\,\alpha \in \mu\,f x\}$. Although the topology can be generalized to the case in which X is *not* locally compact, it is no longer as elegant. In purely topological settings we shall only use it when we have local compactness.

15. Topologies of Uniform Convergence Over Subsets

Let X be a set, Y be a metric space, $\boldsymbol{F(X,Y)}$ denote the set of functions $X \to Y$ (often denoted alternatively by Y^X) and $\mathscr{I} \subseteq \mathscr{P}X$ be an ideal base on X. For each $I \in \mathscr{I}$ we have a gauge d_I on $F(X,Y)$ defined by $d_I(f,g) = \bigvee_{x \in I} d(f\,x, g\,x)$, whose topology we'll denote by $\mathcal{O}_I$, the monads being given by, for $f \in F(X,Y)$ and $g \in F(X,Y)^*$,

$$g \in \mu_I f \Leftrightarrow d_I(f,g) \approx 0 \Leftrightarrow \forall \alpha \in I^*\ g\,\alpha \approx f\,\alpha\,.$$

The conjunction of these topologies $\mathcal{O}_I$ is the *topology of uniform convergence over elements of* $\mathcal{J}$, or the $\mathcal{J}$-uniform topology for short, and the monads are given by, for $f \in F(X,Y)$ and $g \in F(X,Y)^*$,

$$g \in \mu f \Leftrightarrow \forall \alpha \in \nu\mathcal{J} \;\; g\,\alpha \approx f\,\alpha ,$$

where $\nu\mathcal{J}$ recall is the union monad $\bigcup\{I^* \mid I \in \mathcal{J}\}$ of $\mathcal{J}$. In the case where X is a metric space and $\mathcal{J} = Cp\,X$ we have the compact-uniform topology, whose monads are thus given by $g \in \mu f \Leftrightarrow \forall \alpha \in cp\,X^* \;\; g\,\alpha \approx f\,\alpha$, and whose restriction to $C(X,Y)$ is actually the compact-open topology. In the case where X is a metric space and $\mathcal{J} = Bd\,X$ we have the bounded-uniform topology, whose monads are thus given by $g \in \mu f \Leftrightarrow \forall \alpha \in bd\,X^* \;\; g\,\alpha \approx f\,\alpha$, and which coincides of course with the compact-uniform topology if X is boundedly compact. In the case where $\mathcal{J} = \mathcal{P}X$ we have the uniform topology, whose monads are given by $g \in \mu f \Leftrightarrow \forall \alpha \in X^* \;\; g\,\alpha \approx f\,\alpha$. In the case where $\mathcal{J}$ consists of the *finite* subsets of X, we just have the product topology on Y^X, noting that $\nu\mathcal{J} = \{x^* \mid x \in X\}$; this is also known as the *topology of pointwise convergence.* As a general remark, note that the larger $\mathcal{J}$ is, the finer the corresponding $\mathcal{J}$-uniform topology.

16. Uniform Continuity

For metric spaces X and Y we say $g: X^* \to Y^*$ is microcontinuous if $\forall \alpha,\beta \in X^*$, $\alpha \approx \beta \Rightarrow g\,\alpha \approx g\,\beta$; i.e. g preserves $\approx$. Then $f: X \to Y$ is uniformly continuous iff f^* is microcontinuous.

17. Lipschitz Maps and Contractions

Recall that the *Lipschitz ratio* of $f: X \to X$, which we'll denote by r_f, is the least $r \in [0,\infty]$ such that $\forall x,y \in X \;\, d(fx,fy) \le r\, d(x,y)$, and that if $r_f < 1$ f is called a *contraction* of X. Banach's contraction mapping theorem states that any contraction f of a nonempty complete metric space has a unique fixed point (which we'll denote by **fix** f) and for all $x \in X$, $\langle f^n x\rangle \to \text{fix}\, f$. We denote the set of contractions of X by ***Contrac*** X. Letting Lip_1X denote the set of maps $X \to X$ with Lipschitz ratio ≤ 1, it is easily shown that on this set the compact-uniform topology and the topology of pointwise convergence coincide (using microcontinuity of each $g \in Lip_1X^*$), the monads being agreeably given by $\mu f = \{g \in Lip_1X^* \mid \forall x \in X \; \forall \alpha \in \mu x \;\; g\,\alpha \in \mu\, f x\}$; furthermore the binary operation $\circ$ of composition is continuous, so Lip_1X forms a topological monoid.

18. Regular Sets, and Bodies

Recall that a closed subset A of a space X is said to be *regular* if it's the closure of its interior, whilst an open subset V of X is said to be *regular* if it's the interior of its closure. The regular closed sets correspond bijectively with the regular open sets via the mutually inverse maps $A \mapsto \underline{A}$ and $V \mapsto \overline{V}$. The interior of any closed set is a regular open set, whilst the closure of any open set is a regular closed set. For convenience later on, we define (assuming that X is Hausdorff, so every compact set is closed) that a **body** is a regular nonempty compact set, whilst a **body-interior** is an interior of a body, namely a regular nonempty subcompact open set.

19. Residual Closed Sets

Using terminology from [Du], we say a closed set $A \subseteq X$ is **residual** if it has empty interior, equivalently if $A = \partial A$.

Chapter 1

Nonstandard Development of the Vietoris Topology

Section **Page**

0. Introduction

Given a topological space X, the natural idea arises of putting a topology on a collection of subsets of X, a topology defined in terms of that on X. Spaces arising in this way are generally known as 'hyperspaces'. The study of hyperspaces originated with Hausdorff who considered the related *metric* idea, defining a metric on the set of nonempty closed subsets of a bounded metric space X in terms of the original metric on X, and we shall be considering the 'Hausdorff metric' in Chapter 2. But in [Vi], Vietoris considered the purely *topological* situation, defining a topology on the set of closed subsets of a topological space X, and it is this which forms the subject matter of the present chapter, though we'll largely be concentrating on its restriction to the nonempty *compact* subsets of X. This has import for the subject of the Hausdorff metric since, as we'll see in Chapter 2, the topology given by the Hausdorff metric on the nonempty compact subsets of a metric space is in fact the Vietoris topology arising from the topology of X.

The central aspect of our study is the use of nonstandard analysis throughout, allowing some very simple proofs of various well-known standard results. Indeed, the demonstration of the great simplifying power of nonstandard analysis in the study of hyperspaces is meant to be as much a *part* of the work as the actual standard results presented (not *all* of which are in the literature I believe). Juhász determined the monads of the Vietoris topology in [Ju] and gave some applications, and Wattenberg includes a short and likewise nonstandard section on the subject in [Wa], but I know of no *systematic* account of the subject from the nonstandard viewpoint. Whilst the following work is likewise far from comprehensive, it at least deals with a number of basic matters which as far as I know have not appeared in the nonstandard literature before. Indeed, the above two papers are the only nonstandard sources on the Vietoris topology of which I'm aware. The main *standard* references to the Vietoris topology (respectively called the 'finite' topology and the 'exponential' topology in the first two sources) are the paper [Mi] of Ernest Michael, the two-volume work [Ku] on topology by Kuratowski, and the original paper [Vi]. The topology also makes an appearance in [Mat] under the name of the 'myope' topology. To save continual mention, let me state here that many of the standard results included in the following can be found in at least one of the three main sources above ; usually [Mi] and frequently also [Ku].

Throughout, X will denote a Hausdorff space, with $\mathscr{C}X$ denoting the set of closed subsets of X and $\mathscr{K}X$ the set of nonempty compact subsets of X. The nonstandard notation and formulations used are given in 'Preliminaries'.

Finally, here is a summary of the work covered. In Section 1 the Vietoris topology on $\mathscr{C}X$ is defined and the monads given, along with a few basic facts such as X being naturally embedded in $\mathscr{C}X$. We also point out the link beteeen the Vietoris topology and the nonstandard concept of intermediaries. Sections 2 onwards consider the subspace $\mathscr{K}X$, whose properties relate quite strongly to those of X, the monads and a few basic facts being given in Section 2. Section 3 gives a simple nonstandard proof that $\mathscr{K}X$ is compact iff X is, and Section 4 provides a few results on subspaces of $\mathscr{K}X$, showing that for $A \subseteq X$, $\mathscr{K}A$ is dense in $\mathscr{K}X$ iff A is dense in X, and that as long as X is regular, the set of *connected* elements of $\mathscr{K}X$ is closed in $\mathscr{K}X$. Section 5 concerns the fact that the union of any compact subset of $\mathscr{K}X$ is compact, and Section 6 shows how each ideal on X naturally induces an ideal on $\mathscr{K}X$, which is applied in Section 7 to show that $\mathscr{K}X$ is locally compact iff X is. Section 8 concerns maps $\mathscr{K}X \to \mathscr{K}Y$ naturally induced by maps $X \to Y$, and finally, with the foundations of Hutchinson's invariant sets in mind (studied in Chapter 3), Section 9 deals with the idea of taking (with respect to the compact-open topology) a compact set F of continuous maps $Z \to \mathscr{K}Y$ (for Hausdorff Z and Y with Z locally compact) and forming another continuous map $\bigcup F : Z \to \mathscr{K}Y$ by defining $(\bigcup F)z = \bigcup\{fz \mid f \in F\}$. In particular this operation $F \mapsto \bigcup F$ is shown to be continuous.

1. The Vietoris Topology on $\mathscr{C}X$

The open-intersecting topology on $\mathscr{C}X$ is the one generated by the sets of the form **int** $V = \{A \in \mathscr{C}X \mid A \text{ intersects } V\}$ where V is open.

The sub-open topology on $\mathscr{C}X$ is the one generated by the sets of the form **sub** $V = \{A \in \mathscr{C}X \mid A \subseteq V\}$ where V is open. Since these are equivalently the sets of the form **disj** $C = \{A \in \mathscr{C}X \mid A \text{ is disjoint from } C\}$ where C is closed, it may also be known as the closed-avoiding topology on $\mathscr{C}X$.

The Vietoris topology on $\mathscr{C}X$ is the conjunction of the above two topologies. The formulation (3) below of the monads was given by Juhász in [Ju]. Recall that $\mu_X A$ denotes the monad $\bigcap\{N^* \mid N \text{ is a neighbourhood of } A\}$ of A in X.

1.1 Proposition

(1) The monads of the open-intersecting topology on $\mathscr{C}X$ are given by
$$\mu A = \{B \in \mathscr{C}X^* \mid A \subseteq st\, B\}.$$
(2) The monads of the sub-open topology on $\mathscr{C}X$ are given by
$$\mu A = \{B \in \mathscr{C}X^* \mid B \subseteq \mu_X A\}.$$
(3) Hence the monads of the Vietoris topology on $\mathscr{C}X$ are given by
$$\mu A = \{B \in \mathscr{C}X^* \mid A \subseteq st\, B \text{ and } B \subseteq \mu_X A\}.$$

Proof:

(1) We show that for $B \in \mathscr{C}X^*$, $B \in \mu A \Leftrightarrow A \subseteq st\, B$.

$\Rightarrow$: Let $a \in A$. For every open neighbourhood V of a, $A \in \text{int } V$ so $B \in (\text{int } V)^*$, so B intersects V^*. By saturation then, B intersects μa, so $a \in st\, B$.

$\Leftarrow$: For any V with $A \in \text{int } V$ there's $a \in A \cap V$ and $b \in B$ with $b \in \mu a$, so as $\mu a \subseteq V^*$ we have $b \in V^*$ so B intersects V^*, so $B \in (\text{int } V)^*$.

(2) For $B \in \mathscr{C}X^*$, $B \in \mu A \Leftrightarrow \forall$ open V with $A \in \text{sub } V$, $B \in (\text{sub } V)^*$
$\Leftrightarrow \forall$ open V with $A \subseteq V$, $B \subseteq V^*$
$\Leftrightarrow B \subseteq \mu_X A$.

(3) By (1) and (2), taking the intersection of the two monads. ○

For the rest of this section assume $\mathscr{C}X$ has the Vietoris topology. Note that $\emptyset$ is isolated in $\mathscr{C}X$ since $\mu\emptyset = \{\emptyset\}$ by (2) above (and the fact that $\mu_X\emptyset = \emptyset$); hence $\mathscr{C}X - \{\emptyset\}$ is closed in $\mathscr{C}X$. And X is naturally embedded in $\mathscr{C}X$ by $x \mapsto \{x\}$ as shown in the following ;

1.2 Proposition

(1) For $x \in X$, $\mu\{x\}$ consists of the nonempty $*$-closed subsets of μx.
(2) Hence X is embedded in $\mathscr{C}X$ by $x \mapsto \{x\}$.
(3) Considering X as a subspace of $\mathscr{C}X$, X is closed in $\mathscr{C}X$.

Proof:
(1) is easily seen from the last proposition, and it follows that for $\alpha \in X^*$, $\{\alpha\} \in \mu\{x\} \Leftrightarrow \{\alpha\} \subseteq \mu x \Leftrightarrow \alpha \in \mu x$, so (2) holds. Finally, the embedded image of X is closed in $\mathscr{C}X$ since for any $A \in \mathscr{C}X$ and $\alpha \in X^*$ with $\{\alpha\} \in \mu A$, $A \subseteq st\{\alpha\}$ implies (as X is Hausdorff) that A is a singleton. ○

Regarding basic separation properties of $\mathscr{C}X$, we give the following two results ;

<u>1.3</u> <u>Proposition</u> $\mathscr{C}X$ is a T_1 space.
Proof:
Let $C \in \mathscr{C}X$. Then for $A \in \overline{\{C\}}$, $C^* \in \mu A$ so $C^* \subseteq \mu_X A$ giving $C \subseteq X \cap \mu_X A = A$ (the last equality holding as X is T_1), and $A \subseteq st\, C^* = C$. So $A = C$ as required. ○

<u>1.4</u> <u>Proposition</u> The following are equivalent ;

(1) $\mathscr{C}X$ is Hausdorff.
(2) X is regular.
(3) $\forall A \in \mathscr{C}X\ \forall B \in \mathscr{C}X^*,\ B \in \mu A \Rightarrow A = st\, B$.

Proof:
(1) ⇒ (2) : If X is *not* regular there's closed A and $x \notin A$ such that $\mu_X A$ and μx intersect at some α. But then letting $B = A^* \cup \{\alpha\}$, $A^* \subseteq B \subseteq \mu_X A$ gives $B \in \mu A$; yet also $B \in \mu(A \cup \{x\})$ since $A \cup \{x\} \subseteq st\, B$ and $B = A^* \cup \{\alpha\} \subseteq \mu_X A \cup \mu x = \mu_X(A \cup \{x\})$ (as in general $\mu_X A_1 \cup \mu_X A_2 = \mu_X(A_1 \cup A_2)$) ; so $\mathscr{C}X$ is not Hausdorff.
(2) ⇒ (3) : If $B \in \mu A$, then from $B \subseteq \mu_X A$ we get $st\, B \subseteq st\, \mu_X A = A$ (the last equality by regularity of X), which together with $A \subseteq st\, B$ gives $A = st\, B$.
(3) ⇒ (1) : $B \in \mu A_1 \cap \mu A_2 \Rightarrow A_1 = st\, B = A_2$. ○

So when X is regular, an element B of $\mathscr{C}X^*$ can be nearstandard only to its *pointwise* standard part $st\, B = \{x \in X \mid \exists b \in B\ b \in \mu x\}$ in X. In other words, if ${}^\circ B$ exists, it equals $st\, B$. It does not follow however that every B *must* be nearstandard (despite the fact that $st\, B \in \mathscr{C}X$); in other words $\mathscr{C}X$ need not be compact. For example, if B is nonempty and every element of B is remote (so $st\, B = \emptyset$) then B is remote. For locally compact X however there *is* a topology on $\mathscr{C}X$ in which for $A \in \mathscr{C}X$ and $B \in \mathscr{C}X^*$, B is near $A \Leftrightarrow A = st\, B$; this is the S-*compact* topology which we'll encounter in Chapter 4, and which is described separately in Appendix 7.

<u>1.5</u> <u>Proposition</u> The finite subsets of X are dense in $\mathscr{C}X$.
Proof:
Let $A \in \mathscr{C}X$. By saturation A has a hyperfinite intermediary I, and since $A^* \subseteq \mu_X A$ we have $A \subseteq I \subseteq \mu_X A$, giving $I \in \mu A$. ○

As a corollary the *nonempty* finite sets are dense in $\mathscr{C}X - \{\emptyset\}$, hence so is $\mathscr{K}X$. It follows that if X is not compact, $\mathscr{K}X$ is not closed in $\mathscr{C}X - \{\emptyset\}$.

Before moving on to concentrate on the subspace $\mathscr{K}X$ for the rest of the chapter, it will be interesting to see what the monads of the Vietoris topology on $\mathscr{C}X$ are when X is a *discrete* space (equivalently we can regard X as a set), in which case $\mathscr{C}X = \mathscr{P}X$. Since each point x is isolated, i.e. $\mu x = \{x\}$, it follows that $ns\ X^* = X$, so for $B \subseteq X^*$, $st\ B = B \cap X$. For $A \in \mathscr{P}X$ and $B \in \mathscr{P}X^*$ then,

$$\begin{aligned} B \in \mu A &\Leftrightarrow A \subseteq st\ B \text{ and } B \subseteq \mu_X A \\ &\Leftrightarrow A \subseteq B \cap X \text{ and } B \subseteq A^* \qquad\qquad \text{noting } \mu_X A = A^* \\ &\Leftrightarrow A \subseteq B \subseteq A^* \\ &\Leftrightarrow B \text{ is an intermediary of } A\,. \end{aligned}$$

Thus μA consists of the intermediaries of A, providing an interesting link between the latter basic nonstandard concept and the Vietoris topology. Note in particular that the nonstandard result that every subset of X has a hyperfinite intermediary equivalently states that with respect to the Vietoris topology, the finite sets are dense in $\mathscr{C}X$. We saw in 1.5 that the latter is true for *all* X of course.

2. The Vietoris Topology on $\mathscr{K}X$

The **Vietoris topology on $\mathscr{K}X$** is the topology $\mathscr{K}X$ inherits as a subspace of $\mathscr{C}X$ with the Vietoris topology, and for the rest of the chapter $\mathscr{K}X$ is assumed to be endowed with it. The compactness of its elements makes the properties of $\mathscr{K}X$ relate much more closely to those of X than is the case with $\mathscr{C}X$, the monads being given by . . .

2.1 Proposition For $A \in \mathscr{K}X$ and $B \in \mathscr{K}X^*$,

$$\begin{aligned} B \in \mu A &\Leftrightarrow A \subseteq st\, B \text{ and } B \subseteq A^{\mu} \\ &\Leftrightarrow \forall a \in A\ \exists b \in B\ b \in \mu a \text{ and } \forall b \in B\ \exists a \in A\ b \in \mu a \\ &\Leftrightarrow B \subseteq ns\, X^* \text{ and } A = st\, B\,. \end{aligned}$$

Proof:
The first '$\Leftrightarrow$' holds by (3) of 1.1 since as A is compact $\mu_X A = A^{\mu}$. The second statement on the right ("Every element of A has an element of B near it and every element of B is near an element of A") is just a simple rephrasing of the first, and the equivalence of the *third* statement to the first is easily seen. ○

Note that by the above result, even *without* requiring regularity of X (unlike the case for $\mathscr{C}X$), an element B of $\mathscr{K}X^*$ can be nearstandard only to its *pointwise* standard part $st\, B$, and as a consequence . . .

2.2 Proposition $\mathscr{K}X$ is Hausdorff.
Proof:
$B \in \mu A_1 \cap \mu A_2 \Rightarrow A_1 = st\, B = A_2$. ○

Since singletons are compact, we also have by (2) and (3) of 1.2 that . . .

2.3 Note X is embedded in $\mathscr{K}X$ by $x \mapsto \{x\}$, and considered thus as a subspace of $\mathscr{K}X$, X is closed in $\mathscr{K}X$. ○

As implied above, we often identify X with its embedded image in $\mathscr{K}X$.

3. Compactness of $\mathcal{K}X$

The nonstandard formulation that a space Y is compact iff every element of Y^* is nearstandard, allows a simple proof of the fact that $\mathcal{K}X$ is compact iff X is. We need only the following in preparation ;

3.1 Proposition For regular X, and $B \in \mathcal{K}X^*$, $B \in ns\,\mathcal{K}X^* \Leftrightarrow B \subseteq ns\,X^*$.

Proof:

$\Rightarrow$: By 2.1.

$\Leftarrow$: Then $st\,B \in \mathcal{K}X$ (using the background result that for regular X the standard part of any internal set of nearstandard points is compact), and by 2.1, $B \in \mu\; st\,B$. ○

And we now have that . . .

3.2 Proposition $\mathcal{K}X$ is compact $\Leftrightarrow$ X is compact.

Proof:

$\Rightarrow$: As X is closed in $\mathcal{K}X$.

$\Leftarrow$: Then $\forall B \in \mathcal{K}X^*$, $B \subseteq X^* = ns\,X^*$, so as X is also regular, $B \in ns\,\mathcal{K}X^*$. ○

A nonstandard proof of the main implication '$\Leftarrow$' above also appears in [Ju].

4. Subspaces of $\mathcal{K}X$

Any subset A of X forms a topological space under the subspace topology, and we can give the set $\mathcal{K}A$ of nonempty compact subsets of A the Vietoris topology. But $\mathcal{K}A$ is also a subset of $\mathcal{K}X$ and could be given the topology inherited from that on $\mathcal{K}X$. The following shows that these topologies coincide ;

4.1 Note For $A \subseteq X$, the Vietoris topology on $\mathcal{K}A$ is the same as the topology $\mathcal{K}A$ inherits from the Vietoris topology on $\mathcal{K}X$.

Proof:

Giving $\mathcal{K}A$ the Vietoris topology, then for $C \in \mathcal{K}A$ and $B \in \mathcal{K}A^*$, where μ_A denotes 'monad in A' (see 'Subspaces' in 'Preliminaries'),

$$\begin{aligned} B \text{ is near } C^* \text{ in } \mathcal{K}A^* &\Leftrightarrow \forall c \in C\ \exists b \in B\ b \in \mu_A c \text{ and } \forall b \in B\ \exists c \in C\ b \in \mu_A c \\ &\Leftrightarrow \forall c \in C\ \exists b \in B\ b \in \mu c \text{ and } \forall b \in B\ \exists c \in C\ b \in \mu c \\ &\Leftrightarrow B \text{ is near } C^* \text{ in } \mathcal{K}X^*. \end{aligned}$$ ○

And the following gives the criterion for $\mathcal{K}A$ to be dense in $\mathcal{K}X$, using that the set $\mathcal{F}X$ of nonempty finite subsets of X is dense in $\mathcal{K}X$, being dense in $\mathcal{C}X - \{\emptyset\}$;

4.2 Proposition For $A \subseteq X$, $\mathcal{K}A$ is dense in $\mathcal{K}X \Leftrightarrow A$ is dense in X.

Proof:

$\Rightarrow$: For all $x \in X$, there's $B \in \mathcal{K}A^*$ with $B \in \mu\{x\}$, so for some $b \in B \subseteq A^*$, $b \in \mu x$.

$\Leftarrow$: Since $\mathcal{F}X$ is dense in $\mathcal{K}X$ it will suffice to show that $\mathcal{F}A$ is dense in $\mathcal{F}X$. Given $\{x_1, \ldots, x_n\} \in \mathcal{F}X$, for each x_i there's $a_i \in A^*$ with $a_i \in \mu x_i$, in which case $\{a_1, \ldots, a_n\} \in \mu\{x_1, \ldots, x_n\}$, with $\{a_1, \ldots, a_n\} \in \mathcal{F}A^*$. ○

Finally, we give a result on the subspace of *connected* nonempty compact sets ;

4.3 Proposition For regular X, the set of connected elements of $\mathcal{K}X$ is closed in $\mathcal{K}X$.

Proof:

We must show that if B is $*$-connected and $B \in \mu A$, then A is connected. Suppose for a contradiction that it is not. Partition A into closed (hence compact) A_1 and A_2, and let B_1 and B_2 form the unique partition of B such that $A_i = st\, B_i$ (explicitly, $B_i = B \cap A_i{}^{\mu}$). We show that B_1 and B_2 are $*$-closed, giving the contradiction that B is not $*$-connected. For since A_1 and A_2 are disjoint compact sets in a regular Hausdorff space, they have disjoint closed neighbourhoods N_1 and N_2, and as $A_i{}^{\mu} \subseteq N_i{}^*$ with $N_1{}^*$ and $N_2{}^*$ disjoint, we see that $B_i = B \cap N_i{}^*$, which is $*$-closed as claimed. ○

5. The Union Map $\bigcup : \mathcal{K}\mathcal{K}X \to \mathcal{K}X$

Generalizing the fact that the union of any *finite* subset of $\mathcal{K}X$ is compact,

5.1 Proposition The union of any compact subset of $\mathcal{K}X$ is a compact subset of X.

Proof:

For compact $\mathcal{A} \subseteq \mathcal{K}X$, $\forall B \in \mathcal{A}^*$ $\exists A \in \mathcal{A}$ with $B \in \mu A$ so $B \subseteq A^{\mu} \subseteq (\bigcup \mathcal{A})^{\mu}$. So as $(\bigcup \mathcal{A})^* = \bigcup(\mathcal{A}^*)$, we have $(\bigcup \mathcal{A})^* \subseteq (\bigcup \mathcal{A})^{\mu}$, i.e. $\bigcup \mathcal{A}$ is compact. ○

In particular the union of any *nonempty* compact subset of $\mathcal{K}X$ (i.e. element of $\mathcal{K}\mathcal{K}X$) is a *nonempty* compact subset of X (i.e. element of $\mathcal{K}X$). Thus we have what we'll call the **union map** $\bigcup : \mathcal{K}\mathcal{K}X \to \mathcal{K}X$. And with $\mathcal{K}\mathcal{K}X$ having the Vietoris topology induced from the (Vietoris) topology of $\mathcal{K}X$,

5.2 Proposition The union map is continuous.

Proof:

Let $\mathcal{A} \in \mathcal{K}\mathcal{K}X$ and take any $\mathcal{B} \in \mu\mathcal{A}$. We must show that $\bigcup \mathcal{B} \in \mu \bigcup \mathcal{A}$, i.e. $\bigcup \mathcal{A} \subseteq st \bigcup \mathcal{B}$ and $\bigcup \mathcal{B} \subseteq (\bigcup \mathcal{A})^{\mu}$;

(1) $\forall A \in \mathcal{A}$ $\exists B \in \mathcal{B}$ with $B \in \mu A$ so $A \subseteq st\, B$. Thus $\bigcup \mathcal{A} \subseteq \bigcup_{B \in \mathcal{B}} st\, B = st \bigcup \mathcal{B}$.

(2) $\forall B \in \mathcal{B}$ $\exists A \in \mathcal{A}$ with $B \in \mu A$ so $B \subseteq A^{\mu}$. Thus $\bigcup \mathcal{B} \subseteq \bigcup_{A \in \mathcal{A}} A^{\mu} = (\bigcup \mathcal{A})^{\mu}$. ○

5.3 Corollary The n-ary union map $\cup : (\mathcal{K}X)^n \to \mathcal{K}X$ is continuous, giving $(\mathcal{K}X)^n$ the product topology.

Proof:

If $(B_1, \ldots, B_n) \in \mu(A_1, \ldots, A_n)$ then $\forall i$ $B_i \in \mu A_i$, so in $\mathcal{K}\mathcal{K}X$ $\{B_1, \ldots, B_n\} \in \mu\{A_1, \ldots, A_n\}$, so $\bigcup\{B_1, \ldots, B_n\} \in \mu \bigcup\{A_1, \ldots, A_n\}$. ○

6. Induced Ideals on $\mathscr{K}X$

In this section we show how an ideal $\mathscr{I}$ on X induces an ideal $\mathscr{I}_{\mathscr{K}}$ on $\mathscr{K}X$. This is of relevance for example in the next section when we consider the ideal of subcompact sets. Define $\mathscr{I}_{\mathscr{K}} = \{\mathscr{A} \subseteq \mathscr{K}X \mid \bigcup \mathscr{A} \in \mathscr{I}\}$. Then, recalling for (3) below that ν denotes 'union monad of ', we find that . . .

6.1 Proposition For any ideal $\mathscr{I}$ on X,

(1) $\mathscr{I}_{\mathscr{K}}$ is an ideal on $\mathscr{K}X$.

(2) For any basis $\mathscr{B}$ for $\mathscr{I}$, $\{\text{sub } B \mid B \in \mathscr{B}\}$ is a basis for $\mathscr{I}_{\mathscr{K}}$, where $\text{sub } B = \{A \in \mathscr{K}X \mid A \subseteq B\}$.

(3) For $A \in \mathscr{K}X^*$, $A \in \nu\, \mathscr{I}_{\mathscr{K}} \Leftrightarrow A \subseteq \nu \mathscr{I}$.

Proof:

(1) Firstly, if $\mathscr{B} \subseteq \mathscr{A} \in \mathscr{I}_{\mathscr{K}}$ then $\mathscr{B} \in \mathscr{I}_{\mathscr{K}}$ since $\bigcup \mathscr{B} \subseteq \bigcup \mathscr{A} \in \mathscr{I}$ gives $\bigcup \mathscr{B} \in \mathscr{I}$. Secondly, if $\mathscr{A}, \mathscr{B} \in \mathscr{I}_{\mathscr{K}}$ then $\mathscr{A} \cup \mathscr{B} \in \mathscr{I}_{\mathscr{K}}$ since $\bigcup(\mathscr{A} \cup \mathscr{B}) = (\bigcup \mathscr{A}) \cup (\bigcup \mathscr{B}) \in \mathscr{I}$. Lastly, $\emptyset \in \mathscr{I}_{\mathscr{K}}$ since $\bigcup \emptyset = \emptyset \in \mathscr{I}$.

(2) For $B \in \mathscr{B}$, $\text{sub } B \in \mathscr{I}_{\mathscr{K}}$ as $\bigcup \text{ sub } B = B \in \mathscr{I}$. And for all $\mathscr{A} \in \mathscr{I}_{\mathscr{K}}$, as $\bigcup \mathscr{A} \in \mathscr{I}$ there's $B \in \mathscr{B}$ with $\bigcup \mathscr{A} \subseteq B$ so $\mathscr{A} \subseteq \text{sub } B$. Lastly, for all $B_1, B_2 \in \mathscr{B}$ there's $B_3 \in \mathscr{B}$ with $B_1, B_2 \subseteq B_3$, hence $\text{sub } B_1, \text{sub } B_2 \subseteq \text{sub } B_3$.

(3) If $A \in \nu\, \mathscr{I}_{\mathscr{K}}$ there's $\mathscr{A} \in \mathscr{I}_{\mathscr{K}}$ with $A \in \mathscr{A}^*$, so $A \subseteq \bigcup(\mathscr{A}^*) = (\bigcup \mathscr{A})^* \subseteq \nu \mathscr{I}$ as $\bigcup \mathscr{A} \in \mathscr{I}$. Conversely, if $A \subseteq \nu \mathscr{I}$ then for some $B \in \mathscr{I}$, $A \subseteq B^*$ so $A \in (\text{sub } B)^*$ with $\text{sub } B \in \mathscr{I}_{\mathscr{K}}$, so $A \in \nu\, \mathscr{I}_{\mathscr{K}}$. ○

$\mathscr{I}_{\mathscr{K}}$ is the <u>ideal on $\mathscr{K}X$ induced by $\mathscr{I}$</u>, and since ideals are *determined* by their union monads note that (3) characterizes $\mathscr{I}_{\mathscr{K}}$ in that for any ideal $\mathscr{J}$ on $\mathscr{K}X$, $\mathscr{J} = \mathscr{I}_{\mathscr{K}}$ iff for all $A \in \mathscr{K}X^*$, $A \in \nu \mathscr{J} \Leftrightarrow A \subseteq \nu \mathscr{I}$. This characterization is used in the next section to show that if $\mathscr{I}$ is the ideal of subcompact subsets of X then $\mathscr{I}_{\mathscr{K}}$ is the ideal of subcompact subsets of $\mathscr{K}X$. The *following* concept will be used in Chapter 2. We'll say an ideal $\mathscr{I}$ on a space X is <u>**topological**</u> if every element of X has a neighbourhood belonging to $\mathscr{I}$. The nonstandard formulation is that . . .

6.2 Note For an ideal $\mathscr{I}$ on X, $\mathscr{I}$ is topological $\Leftrightarrow$ $ns\, X^* \subseteq \nu \mathscr{I}$.

Proof:

$$\begin{aligned} \mathscr{I} \text{ is topological } &\Leftrightarrow \forall x \in X, \text{ some element of } \mathscr{N}x \text{ belongs to } \mathscr{I} \\ &\Leftrightarrow \forall x \in X, \ \mu \mathscr{N}x \subseteq \nu \mathscr{I} \qquad \text{(by a result on monads)} \\ &\Leftrightarrow \forall x \in X, \ \mu x \subseteq \nu \mathscr{I} \\ &\Leftrightarrow ns\, X^* \subseteq \nu \mathscr{I}. \end{aligned}$$

○

We can use this to prove that . . .

<u>**6.3**</u> <u>**Proposition**</u> For any ideal $\mathcal{J}$ on X,

$$\mathcal{J}_{\mathcal{K}} \text{ is topological} \Leftrightarrow \mathcal{J} \text{ is topological.}$$

Proof:

$\Rightarrow$: For $\alpha \in \text{ns}\, X^*$, as $\{\alpha\} \in ns\mathcal{K}X^*$ we have $\{\alpha\} \in \nu\, \mathcal{J}_{\mathcal{K}}$, so $\{\alpha\} \subseteq \nu\mathcal{J}$, so $\alpha \in \nu\mathcal{J}$.

$\Leftarrow$: For $B \in ns\mathcal{K}X^*$, $B \in \nu\, \mathcal{J}_{\mathcal{K}}$ since $B \subseteq ns\, X^* \subseteq \nu\mathcal{J}$. ○

7. Local Compactness of $\mathcal{K}X$

Recall that for a Hausdorff space Y the set $cp\ Y^*$ of *compact* points of Y^* is the union monad of the ideal $subCp\ Y$ of subcompact subsets of Y, and that Y is locally compact iff $ns\ Y^* \subseteq cp\ Y^*$ (i.e., in the terminology of the last section, $subCp\ Y$ is topological). We have the following criterion for the subcompact subsets of $\mathcal{K}X$, equivalently for the compact elements of $\mathcal{K}X^*$;

7.1 Proposition $subCp\,\mathcal{K}X = (subCp\ X)_{\mathcal{K}}$,

i.e. for $\mathcal{A} \subseteq \mathcal{K}X$, $\mathcal{A}$ is subcompact $\Leftrightarrow$ $\bigcup \mathcal{A}$ is subcompact in X,

i.e. for $A \in \mathcal{K}X^*$, $A \in cp\,\mathcal{K}X^* \Leftrightarrow A \subseteq cp\ X^*$.

Proof:

We prove the last line, which as explained in the previous section, amounts to the first.

$\Rightarrow$: Where $A \in \mathcal{A}^*$ with $\mathcal{A}$ a compact subset of $\mathcal{K}X$, $A \subseteq \bigcup(\mathcal{A}^*) = (\bigcup \mathcal{A})^* \subseteq cp\ X^*$ as $\bigcup \mathcal{A}$ is compact (by 5.1).

$\Leftarrow$: Then as A is internal, $A \subseteq C^*$ for some compact $C \subseteq X$. Thus $A \in \mathcal{K}C^*$ (since $\mathcal{K}C^*$ consists of the nonempty $*$-compact subsets of C^*) and as C is compact so is $\mathcal{K}C$, which is a subset of $\mathcal{K}X$, so $A \in cp\,\mathcal{K}X^*$ as required. o

We can now give a short nonstandard proof the following well-known result ;

7.2 Proposition $\mathcal{K}X$ is locally compact $\Leftrightarrow$ X is locally compact.

Proof:

$\Rightarrow$: As X is closed in $\mathcal{K}X$.

$\Leftarrow$: Then $\forall A \in cp\,\mathcal{K}X^*$, $A \subseteq cp\ X^* = ns\ X^*$ so by 3.1 $A \in ns\,\mathcal{K}X^*$. o

The *standard* proof corresponding to the above nonstandard one, made into a chain of 'iffs' for the sake of elegance, is as follows ;

$\mathcal{K}X$ is locally compact $\Leftrightarrow$ $subCp\,\mathcal{K}X$ is topological

$\Leftrightarrow$ $(subCp\ X)_{\mathcal{K}}$ is topological

$\Leftrightarrow$ $subCp\ X$ is topological (by 6.3)

$\Leftrightarrow$ X is locally compact.

8. The Map $C(X,Y) \to C(\mathcal{K}X, \mathcal{K}Y)$

Throughout let X and Y be Hausdorff spaces with $\mathcal{K}X$ and $\mathcal{K}Y$ both given the Vietoris topology. We first note that continuous maps $X \to Y$ naturally induce continuous maps $\mathcal{K}X \to \mathcal{K}Y$ as follows ;

8.1 Proposition If $f: X \to Y$ is continuous, so is the naturally induced map $f_{\mathcal{K}}: \mathcal{K}X \to \mathcal{K}Y$ defined by $f_{\mathcal{K}} A = fA = \{fa \mid a \in A\}$.

Proof:

Note that as f is continuous it preserves compactness, so $f_{\mathcal{K}}$ *does* map $\mathcal{K}X$ into $\mathcal{K}Y$. Now we prove continuity. If $B \in \mu A$ then in other words $\forall b \in B\ \exists a \in A\ b \in \mu a$ and $\forall a \in A\ \exists b \in B\ b \in \mu a$; but continuity of f gives that $b \in \mu a \Rightarrow fb \in \mu fa$; so $fB \in \mu fA$, i.e. $f_{\mathcal{K}} B \in \mu f_{\mathcal{K}} A$ as required. ○

We thus have a map $_{\mathcal{K}}: C(X,Y) \to C(\mathcal{K}X, \mathcal{K}Y)$ from the set of continuous maps $X \to Y$ into the set of continuous maps $\mathcal{K}X \to \mathcal{K}Y$. Assuming now that X is locally compact (so $\mathcal{K}X$ is too), and giving $C(X,Y)$ and $C(\mathcal{K}X, \mathcal{K}Y)$ the compact-open topology, then . . .

8.2 Proposition $_{\mathcal{K}}: C(X,Y) \to C(\mathcal{K}X, \mathcal{K}Y)$ is continuous.

Proof:

Let $f \in C(X,Y)$ and $g \in \mu f$. We must show that $g_{\mathcal{K}} \in \mu f_{\mathcal{K}}$, i.e. that for all $A \in \mathcal{K}X$ and $B \in \mathcal{K}X^*$ with $B \in \mu A$, $g_{\mathcal{K}} B \in \mu f_{\mathcal{K}} A$. Whenever $a \in A$ and $b \in B$ with $b \in \mu a$ we have $g\,b \in \mu\, f\,a$. So since every element of A has an element of B near it and every element of B is near an element of A, the same is true of fA and gB, so $gB \in \mu fA$ as required. ○

Note also that if $f: X \cong Y$ then $f_{\mathcal{K}}: \mathcal{K}X \cong \mathcal{K}Y$ with $f_{\mathcal{K}}^{-1} = f^{-1}{}_{\mathcal{K}}$.

9. Union Functions

Letting Z be a locally compact Hausdorff space throughout, with $C(Z,\mathcal{K}X)$ having the compact-open topology and $\mathcal{K}C(Z,\mathcal{K}X)$ the induced Vietoris topology, in this section we show how a nonempty compact set F of continuous maps $Z \to \mathcal{K}X$ gives *another* continuous map $\bigcup F : Z \to \mathcal{K}X$ defined by $(\bigcup F)z = \bigcup_{f \in F} fz$. This leads on to the matter of forming from a nonempty compact set F of continuous operations on X a continuous operation $\bigcup F$ on $\mathcal{K}X$ by $(\bigcup F)A = \bigcup_{f \in F} fA$, which is of relevance to Chapter 3 in forming the foundations of Hutchinson's invariant sets.

For $F \subseteq C(Z,\mathcal{K}X)$ and $z \in Z$ we define $\boldsymbol{Fz} = \{fz \mid f \in F\}$. Then . . .

9.1 Note For $F \in \mathcal{K}C(Z,\mathcal{K}X)$ and $z \in Z$, $Fz \in \mathcal{K}\mathcal{K}X$ so $\bigcup(Fz) \in \mathcal{K}X$.
Proof:
Fz is the image of F under the continuous map $C(Z,\mathcal{K}X) \to \mathcal{K}X$ in which $f \mapsto fz$. Hence as F is compact so is Fz; hence so is the union (by 5.1). ○

So for $F \in \mathcal{K}C(Z,\mathcal{K}X)$ we define $\bigcup F : Z \to \mathcal{K}X$ by $(\bigcup F)z = \bigcup(Fz) = \bigcup_{f \in F} fz$, which we'll call the **union function** of F. As the following shows, $\bigcup F$ is continuous ;

9.2 Proposition

(1) The map $\mathcal{K}C(Z,\mathcal{K}X) \times Z \to \mathcal{K}\mathcal{K}X$ in which $(F,z) \mapsto Fz$, is continuous.
(2) The map $\mathcal{K}C(Z,\mathcal{K}X) \times Z \to \mathcal{K}X$ in which $(F,z) \mapsto (\bigcup F)z$, is continuous.
(3) For $F \in \mathcal{K}C(Z,\mathcal{K}X)$, $\bigcup F$ is continuous.

Proof:
(1) Let $(F,z) \in \mathcal{K}C(Z,\mathcal{K}X) \times Z$ and take any $(G,\alpha) \in \mu(F,z)$, i.e. $G \in \mu F$ and $\alpha \in \mu z$. We must show that $G\alpha \in \mu\, Fz$, i.e. $\{g\alpha \mid g \in G\} \in \mu\{fz \mid f \in F\}$.
Since $\forall g \in G\ \exists f \in F\ g \in \mu f$ and $\forall f \in F\ \exists g \in G\ g \in \mu f$, it remains only to note that in general if $g \in \mu f$ then $g\alpha \in \mu\, fz$ since Z is locally compact.
(2) By (1), following the map there with the continuous map $\bigcup : \mathcal{K}\mathcal{K}X \to \mathcal{K}X$.
(3) By (2), keeping F fixed. ○

We thus have a map $\bigcup : \mathcal{K}C(Z,\mathcal{K}X) \to C(Z,\mathcal{K}X)$, producing from each nonempty compact set of continuous maps $Z \to \mathcal{K}X$ another continuous map $Z \to \mathcal{K}X$. And by (2) above,

9.3 Proposition $\bigcup : \mathcal{K}C(Z,\mathcal{K}X) \to C(Z,\mathcal{K}X)$ is continuous.
Proof:
Let $F \in \mathcal{K}C(Z,\mathcal{K}X)$ and $G \in \mu F$. Then for all $z \in Z$ and $\alpha \in \mu z$, $(G,\alpha) \in \mu(F,z)$ so $(\bigcup G)\alpha \in \mu(\bigcup F)z$. So $\bigcup G \in \mu \bigcup F$ as required. ○

Assuming now that X is locally compact we can use the above work to show in particular how every nonempty compact set F of continuous maps $X \to X$ induces a continuous operation on $\mathscr{K}X$. For since the map $_{\mathscr{K}} : C(X,X) \to C(\mathscr{K}X, \mathscr{K}X)$ described in the last section is continuous, it maps each $F \in \mathscr{K}C(X,X)$ to an element $F_{\mathscr{K}} = \{f_{\mathscr{K}} \mid f \in F\}$ of $\mathscr{K}C(\mathscr{K}X, \mathscr{K}X)$, from which we in turn (considering $Z = \mathscr{K}X$, being locally compact as X is) get the union function $\bigcup(F_{\mathscr{K}}) : \mathscr{K}X \to \mathscr{K}X$, which we'll naturally call the **union function** of F, denoted by $\bigcup F$ and given by $(\bigcup F)A = \bigcup(F_{\mathscr{K}} A) = \bigcup_{f \in F} fA = FA$ where $FA = \{fa \mid f \in F \text{ and } a \in A\}$. By 9.2(3), $\bigcup(F_{\mathscr{K}})$ (namely $\bigcup F$) is continuous, and we therefore now have a map $\bigcup : \mathscr{K}C(X,X) \to C(\mathscr{K}X, \mathscr{K}X)$ sending each F to $\bigcup F$. And by past results,

9.4 Corollary For locally compact X, $\bigcup : \mathscr{K}C(X,X) \to C(\mathscr{K}X, \mathscr{K}X)$ is continuous.

Proof:

It's the composition of $_{\mathscr{K}} : \mathscr{K}C(X,X) \to \mathscr{K}C(\mathscr{K}X, \mathscr{K}X)$ followed by the map $\bigcup : \mathscr{K}C(\mathscr{K}X, \mathscr{K}X) \to C(\mathscr{K}X, \mathscr{K}X)$; and the latter is continuous by 9.3 applied with $Z = \mathscr{K}X$, whilst the former is continuous as it's the map naturally induced by the continuous map $_{\mathscr{K}} : C(X,X) \to C(\mathscr{K}X, \mathscr{K}X)$ (apply 8.1 with X and Y replaced there by $C(X,X)$ and $C(\mathscr{K}X, \mathscr{K}X)$). ○

In connection with the above corollary, see the last paragraph of Appendix 1 for an alternative perspective.

Chapter 2

Nonstandard Development of the Hausdorff Metric

Section **Page**

0. Introduction

In the last chapter we dealt mainly with the Vietoris topology on the set $\mathcal{K}X$ of nonempty compact subsets of a Hausdorff space X. In *this* chapter we deal with the Hausdorff metric h on $\mathcal{K}X$ for a *metric* space X, which in fact *induces* the Vietoris topology on $\mathcal{K}X$, as is well-known. Originally introduced in [Haus], the Hausdorff metric has come agreeably to the fore a little in recent years in connection with the subject of fractals, more specifically with those of the type considered by Hutchinson and described in Chapter 3, and the present chapter, backed up by the last, develops the theory required. As with the last chapter the simplifying power of nonstandard analysis is used throughout, providing among other things a short proof of the fact that if X is complete so is $\mathcal{K}X$. Since writing this chapter I have received a preprint [Li,1] from Tom Lindstrøm which includes a short section *likewise* considering aspects of the Hausdorff metric in a nonstandard light, independently giving a nonstandard proof of the above completeness result and also (essentially in 2.5) of the fact that if X is boundedly compact so is $\mathcal{K}X$. Other than this I know of no further nonstandard sources. The main *standard* reference of which I'm aware is the two-volume work [Ku] on topology by Kuratowski, which includes a standard proof of the above-mentioned completeness result (credited to Hahn).

Throughout, X denotes a metric space with metric d, and $\mathcal{K}X$ denotes the set of nonempty compact subsets of X. Any nonstandard notation or formulations not explained in the text may be found in 'Preliminaries'.

Lastly, here is an outline of the chapter. In Section 1 the Hausdorff metric h on $\mathcal{K}X$ is defined through the idea of distinguishability; there is a critical level of visual resolution beyond which A and B are distinguishable, and this resolution level is the Hausdorff distance $h(A,B)$ between A and B. We show that h gives the Vietoris topology, and that X is metrically embedded in $\mathcal{K}X$ by $x \mapsto \{x\}$. Section 2 gives a criterion for a subset of $\mathcal{K}X$ to be bounded, and goes on to show that $\mathcal{K}X$ is boundedly compact iff X is. For $X = \mathbb{R}^n$ this is essentially the 'Blaschke Selection Theorem'. The main result in Section 3, and a cornerstone used in Hutchinson's theory when Banach's contraction mapping theorem is called upon, is that if X is complete so is $\mathcal{K}X$. We follow in Section 4 with results on sequences in $\mathcal{K}X$ and their limits. Sections 5 and 6 on functions parallel Sections 8 and 9 of Chapter 1, but use topologies of uniform convergence in general, as opposed to just the compact-open topology (which is the compact-uniform topology, in the metric context), and particular reference is made to Lipschitz maps in preparation for Chapter 3.

1. The Hausdorff Metric on $\mathscr{K}X$

For $A,B \in \mathscr{K}X$ and $\delta \in [0,\infty)$ we'll say A and B are **δ-indistinguishable** if every element of A is within distance δ of some element of B and every element of B is within distance δ of some element of A. You can think of this as meaning that if you cannot resolve points distance $\leq \delta$ apart then A and B are visually indistinguishable. The 'Hausdorff distance' between A and B is going to be the least such δ, hence the critical resolution level beyond which A and B can be distinguished. But first we must show a minimum exists. To this end, note that δ-indistinguishability can be expressed conveniently in terms of 'fringes' as follows. For nonempty $A \subseteq X$ and $\delta \in [0,\infty)$, the **closed δ-fringe** of A is

$$[A]_\delta = \{x \in X \mid d(x,A) \leq \delta\} \quad \text{where } d(x,A) = \bigwedge \{d(x,a) \mid a \in A\}.$$

For $A \in \mathscr{K}X$, $d(x,A) = \min\{d(x,a) \mid a \in A\}$ as the map $a \mapsto d(x,a)$ is continuous on compact A hence achieves a minimum, so $[A]_\delta = \{x \in X \mid \exists\, a \in A \;\; d(x,a) \leq \delta\}$. So for $A,B \in \mathscr{K}X$ note that

$$A \text{ and } B \text{ are } \delta\text{-indistinguishable} \Leftrightarrow A \subseteq [B]_\delta \text{ and } B \subseteq [A]_\delta .$$

Closed fringes are indeed closed, and in general $[A]_0 = \overline{A}$. For $y \in X$, the closed δ-fringe of $\{y\}$ is $\{x \in X \mid d(x,y) \leq \delta\}$, more usually called the **closed δ-ball** on y and denoted by $[y]_\delta$.

1.1 Lemma For all $A,B \in \mathscr{K}X$ there's a least $\delta \in [0,\infty)$ such that $A \subseteq [B]_\delta$.

Proof:
Letting $\Delta = \{\delta \in [0,\infty) \mid A \subseteq [B]_\delta\}$, Δ is nonempty and $A \subseteq \bigcap\{[B]_\delta \mid \delta \in \Delta\} = [B]_{\wedge\Delta}$ where $\wedge\Delta$ is the g.l.b. of Δ, so $\wedge\Delta$ is the least element of Δ. ○

And hence, as required,

1.2 Proposition For all $A,B \in \mathscr{K}X$ there's a least $\delta \in [0,\infty)$ such that A and B are δ-indistinguishable.

Proof:
Namely let $\delta = \max\{\delta_1, \delta_2\}$ where δ_1 and δ_2 are respectively least such that $A \subseteq [B]_{\delta_1}$ and $B \subseteq [A]_{\delta_2}$; then for any $r \in [0,\infty)$,

$$A \subseteq [B]_r \text{ and } B \subseteq [A]_r \Leftrightarrow r \geq \delta_1 \text{ and } r \geq \delta_2 \Leftrightarrow r \geq \delta. \qquad ○$$

The δ of the above proposition is the **Hausdorff distance** between A and B, denoted here by $h(A,B)$.

By definition we have that for all $\delta \geq 0$,

$$h(A,B) \leq \delta \iff A \text{ and } B \text{ are } \delta\text{-indistinguishable.}$$

And we easily find that . . .

1.3 Proposition h is a metric on $\mathscr{K}X$ with (X,d) embedded in $(\mathscr{K}X, h)$ by $x \mapsto \{x\}$.

Proof:

(1) $h(A,B) = 0 \iff A$ and B are 0-indistinguishable
$\iff A \subseteq B$ and $B \subseteq A$
$\iff A = B$.

(2) $h(A,B) = h(B,A)$.

(3) To see $h(A,C) \leq h(A,B) + h(B,C)$ we use the triangle inequality for d. For $a \in A$ there's $b \in B$ within distance $h(A,B)$ of a, and there's $c \in C$ within distance $h(B,C)$ of b, so this c is within distance $h(A,B) + h(B,C)$ of a. Likewise every $c \in C$ is within this distance of some $a \in A$.

(4) Note that $h(\{x\},\{y\}) = d(x,y)$. ○

h is called the **Hausdorff metric** on $\mathscr{K}X$ and for the remainder of the chapter $\mathscr{K}X$ is assumed to be endowed with it, and we'll identify the embedded image of X with X. As we'll be using nonstandard analysis in the study of $\mathscr{K}X$, a natural first step is to express the relation of sets being infinitesimally distant in $\mathscr{K}X^*$ in terms of the relation of points being infinitesimally distant in X^*. To this end we'll say $A,B \subseteq X^*$ are **infinitesimally indistinguishable** if every element of A is infinitesimally distant from some element of B and vice versa. Equivalently, defining the **infinitesimal fringe** of $A \subseteq X^*$ to be $A^{\approx} = \{x \in X^* \mid \exists a \in A \;\; x \approx a\} = \bigcup\{a^{\approx} \mid a \in A\}$ where $a^{\approx}$ denotes $\{x \in X^* \mid x \approx a\}$, note that for $A,B \subseteq X^*$,

$$A \text{ and } B \text{ are infinitesimally indistinguishable} \iff A \subseteq B^{\approx} \text{ and } B \subseteq A^{\approx}$$
$$\iff A^{\approx} = B^{\approx}.$$

For *internal* nonempty $A \subseteq X^*$ incidentally, note that $A^{\approx} = \{x \in X^* \mid d(x,A) \approx 0\}$. We now give the awaited formulation of the relation $\approx$ in $\mathscr{K}X^*$;

1.4 Proposition For $A,B \in \mathscr{K}X^*$,

$$A \approx B \iff A \text{ and } B \text{ are infinitesimally indistinguishable.}$$

Proof:

$\Rightarrow$: Then every $a \in A$ is within distance $h(A,B) \approx 0$ of some $b \in B$ and vice versa .

$\Leftarrow$: Then for all real $\delta > 0$, $h(A,B) \leq \delta$. Hence $h(A,B) \approx 0$ as required. ○

And as a consequence, noting that $A^{\approx} = (A^*)^{\approx}$ for $A \in \mathscr{K}X$, . . .

1.5 Proposition The topology on $\mathcal{K}X$ induced by h is the Vietoris topology.

Proof:

For $A \in \mathcal{K}X$ and $B \in \mathcal{K}X^*$,

$$B \approx A^* \Leftrightarrow B \text{ and } A^* \text{ are infinitesimally indistinguishable}$$
$$\Leftrightarrow B \text{ and } A \text{ are infinitesimally indistinguishable}$$
$$\Leftrightarrow B \in \mu A \text{ in the Vietoris topology (as in general } b \approx a \Leftrightarrow b \in \mu a\text{).} \quad \circ$$

The results of Chapter 1 (Sections 2 onwards) thus apply in the present situation and will be assumed in the following work. A few of the necessarily *topological* results of that chapter can be improved to *metric* results. In particular, in connection with 4.1 of Chapter 1, for any subset A of X, $\mathcal{K}A$ moreover forms a *metric* subspace of $\mathcal{K}X$ under the Hausdorff metric induced from the metric on A. And regarding 5.2 of Chapter 1 it's quite easy to show (using the usual nonstandard formulation) that the union map $\bigcup : \mathcal{K}\mathcal{K}X \to \mathcal{K}X$ is *uniformly* continuous, giving $\mathcal{K}\mathcal{K}X$ the Hausdorff metric induced by the Hausdorff metric h on $\mathcal{K}X$.

We end this section with an inequality on the Hausdorff distance between unions, to be used later ;

1.6 Note $h(\bigcup_i A_i, \bigcup_i B_i) \leq \bigvee_i h(A_i, B_i)$ where $A_i, B_i, \bigcup_i A_i, \bigcup_i B_i$ are all in $\mathcal{K}X$.

Proof:

Each $a \in \bigcup_i A_i$ is in some A_j hence within distance $h(A_j, B_j)$ of some $b \in B_j$ hence within distance $\bigvee_i h(A_i, B_i)$ of this $b \in \bigcup_i B_i$. Likewise vice versa. $\circ$

2. Boundedness and Bounded Compactness of $\mathscr{K}X$

Recalling that we identify the embedded image of X in $\mathscr{K}X$ with X itself, the following note gives the recipe for the Hausdorff distance of an element of $\mathscr{K}X$ from a point x of X. Recall that $[x]_\delta$ denotes the closed δ-ball on x in X.

2.1 Note For $A \in \mathscr{K}X$ and $x \in X$, $h(A,x) \leq \delta \Leftrightarrow A \subseteq [x]_\delta$, so

$$h(A,x) = \bigvee\{d(a,x) \mid a \in A\}.$$

Proof:
The first part is simple, and from this it follows that $h(A,x)$ is the least δ such that $A \subseteq [x]_\delta$, i.e. such that $\forall a \in A \;\; d(a,x) \leq \delta$. ○

As a trivial corollary, in $\mathscr{K}X$ the closed δ-ball on $\{x\}$ consists of the nonempty compact subsets of $[x]_\delta$. Now, recalling that for a metric space Y the set of 'bounded' points of Y^* is $bd\ Y^* = \{\alpha \in Y^* \mid \alpha \text{ is finitely distant from some (hence all) } y \in Y\}$, which is the union monad of the ideal $Bd\ Y$ of bounded subsets of Y, we have the following criterion for the bounded subsets of $\mathscr{K}X$, equivalently (bearing in mind Section 6 of Chapter 1) for the bounded elements of $\mathscr{K}X^*$;

2.2 Proposition $Bd\,\mathscr{K}X = (Bd\ X)_{\mathscr{K}}$,

i.e. for $\mathscr{A} \subseteq \mathscr{K}X$, $\mathscr{A}$ is bounded $\Leftrightarrow \bigcup \mathscr{A}$ is bounded in X,

i.e. for $A \in \mathscr{K}X^*$, $A \in bd\,\mathscr{K}X^* \Leftrightarrow A \subseteq bd\ X^*$.

Proof:
We prove the last line, which as explained in Section 6 of Chapter 1, amounts to the first. Taking any $x \in X$, we have

$$\begin{aligned} A \in bd\,\mathscr{K}X^* &\Leftrightarrow h(A,x) \text{ is finite} \\ &\Leftrightarrow \bigvee\{d(a,x) \mid a \in A\} \text{ is finite} \\ &\Leftrightarrow \forall a \in A \;\; d(a,x) \text{ is finite} \\ &\Leftrightarrow A \subseteq bd\ X^*, \end{aligned}$$

using for the third '$\Leftrightarrow$' that the least upper bound of any internal set of finite hyperreals is finite. ○

And since $\bigcup \mathscr{K}X = X$, we immediately have that . . .

2.3 Corollary $\mathscr{K}X$ is bounded $\Leftrightarrow$ X is bounded. ○

From knowledge of the Vietoris topology we already know that $\mathscr{K}X$ is compact iff X is. The nonstandard formulation of when a metric space Y is *boundedly* compact (namely that every *bounded* element of Y^* be nearstandard) allows the following simple proof of the fact that $\mathscr{K}X$ is boundedly compact iff X is ;

2.4 **Proposition** $\mathscr{K}X$ is boundedly compact $\Leftrightarrow$ X is boundedly compact.

Proof:

$\Rightarrow$: As X is closed in $\mathscr{K}X$.

$\Leftarrow$: Then $\forall B \in bd\mathscr{K}X^*$, $B \subseteq bd\, X^* = ns\, X^*$ so (by 3.1 of Chapter 1) $B \in ns\mathscr{K}X^*$. o

An alternative *standard* proof uses (along with 2.2 above, and 7.1 of Chapter 1) the easily proved result that for ideals $\mathscr{I}$ and $\mathscr{J}$ on X, $\mathscr{I}_{\mathscr{K}} \subseteq \mathscr{J}_{\mathscr{K}} \Leftrightarrow \mathscr{I} \subseteq \mathscr{J}$, as follows ;

$$\begin{aligned} \mathscr{K}X \text{ is boundedly compact } &\Leftrightarrow Bd\mathscr{K}X \subseteq subCp\mathscr{K}X \\ &\Leftrightarrow (Bd\, X)_{\mathscr{K}} \subseteq (subCp\, X)_{\mathscr{K}} \\ &\Leftrightarrow Bd\, X \subseteq subCp\, X \\ &\Leftrightarrow X \text{ is boundedly compact.} \end{aligned}$$

In particular, since $\mathbb{R}^n$ is boundedly compact so is $\mathscr{K}\,\mathbb{R}^n$. This result in fact amounts to the 'Blaschke Selection Theorem' (e.g. see [Fa]) that for any infinite subset $\mathscr{A}$ of $\mathscr{K}\,\mathbb{R}^n$ whose elements lie within some bounded region of $\mathbb{R}^n$ (i.e. $\bigcup \mathscr{A}$ is bounded) there is an element of $\mathscr{K}\,\mathbb{R}^n$ to which some sequence of distinct elements of $\mathscr{A}$ converges with respect to the Hausdorff metric. For this statement just says that every bounded infinite subset of $\mathscr{K}\,\mathbb{R}^n$ has a cluster point; equivalently that $\mathscr{K}\,\mathbb{R}^n$ is boundedly compact.

3. Completeness of $\mathcal{K}X$

For a metric space Y, $\alpha \in Y^*$ is said to be **pre-nearstandard** if for all real $\epsilon > 0$ there's $y \in Y$ with $d(y^*,\alpha) \leq \epsilon$, the terminology arising from the fact that α is pre-nearstandard in Y^* iff α is nearstandard in $(Y^c)^*$ where Y^c denotes 'the' completion of Y (this being unique up to isometry over Y). The set of pre-nearstandard points of Y^* is denoted by $\boldsymbol{pns}\ Y^*$, and we have the nonstandard formulation that Y is complete iff every pre-nearstandard point of Y^* is nearstandard, i.e. $pns\ Y^* = ns\ Y^*$ (see [Da], Theorem 5.20). Noting that . . .

3.1 Lemma For $A \in \mathcal{K}X^*$, $\quad A \in pns\ \mathcal{K}X^* \Rightarrow A \subseteq pns\ X^*$.

Proof:

Let $A \in pns\,\mathcal{K}X^*$ and take any $\alpha \in A$. Then for any real $\epsilon > 0$, where $B \in \mathcal{K}X$ with $h(B^*,A) \leq \epsilon/2$ there's $\beta \in B^*$ with $d(\beta,\alpha) \leq \epsilon/2$, and in turn β is near some $b \in B$, giving $d(b,\alpha) \leq \epsilon$. So α is pre-nearstandard. ○

. . . then we have the following criterion for completeness of $\mathcal{K}X$;

3.2 Proposition $\quad \mathcal{K}X$ is complete $\Leftrightarrow$ X is complete.

Proof:

$\Rightarrow$: As X is closed in $\mathcal{K}X$.

$\Leftarrow$: $A \in pns\,\mathcal{K}X^* \Rightarrow A \subseteq pns\ X^* = ns\ X^* \Rightarrow A \in ns\,\mathcal{K}X^*$. ○

Using this we find that the completion of $\mathcal{K}X$ is $\mathcal{K}(X^c)$;

3.3 Proposition $\quad (\mathcal{K}X)^c = \mathcal{K}(X^c)$

Proof:

X is dense in X^c so by 4.2 of Chapter 1 $\mathcal{K}X$ is dense in $\mathcal{K}(X^c)$, which is complete as X^c is, hence $\mathcal{K}(X^c)$ is a completion of $\mathcal{K}X$. ○

And using also the aforementioned result that $\alpha \in Y^*$ is pre-nearstandard in Y^* iff α is nearstandard in $(Y^c)^*$, we can provide the converse of 3.1 to round off the picture ;

3.4 Proposition For $A \in \mathcal{K}X^*$, $\quad A \in pns\,\mathcal{K}X^* \Leftrightarrow A \subseteq pns\ X^*$.

Proof:

$$\begin{aligned} A \in pns\,\mathcal{K}X^* &\Leftrightarrow A \in ns\,((\mathcal{K}X)^c)^* \\ &\Leftrightarrow A \in ns\,(\mathcal{K}(X^c))^* \qquad \text{as } (\mathcal{K}X)^c = \mathcal{K}(X^c) \\ &\Leftrightarrow A \subseteq ns\,(X^c)^* \\ &\Leftrightarrow A \subseteq pns\ X^*. \end{aligned}$$ ○

4. Limits of Sequences in $\mathcal{K}X$

For a sequence $A = \langle A_n \rangle$ of nonempty subsets of X we'll say x is a limit point of $\langle A_n \rangle$ with respect to d if $\langle d(x,A_n) \rangle \to 0$. We'll call the set of limit points of $\langle A_n \rangle$ the limit set of $\langle A_n \rangle$ with respect to d, denoted by $\mathbf{lim}_d \langle A_n \rangle$; the subscript d is there so as not to confuse with the possible limit $\lim \langle A_n \rangle$ of $\langle A_n \rangle$ in $\mathcal{K}X$ when the sets A_n are in $\mathcal{K}X$.

4.1 Proposition For a sequence $A = \langle A_n \rangle$ of nonempty subsets of X,

(1) $x \in \lim_d \langle A_n \rangle \Leftrightarrow A_n$ intersects μx for all *infinite* n.

(2) $\lim_d \langle A_n \rangle = \bigcap \{st\, A_n \mid n \text{ is infinite}\}$.

(3) $\lim_d \langle A_n \rangle$ is closed.

Proof:

(1) $x \in \lim_d \langle A_n \rangle \Leftrightarrow \langle d(x,A_n) \rangle \to 0$

$\Leftrightarrow \forall$ infinite n $d(x,A_n) \approx 0$

$\Leftrightarrow \forall$ infinite n $x \in A_n^{\approx}$

$\Leftrightarrow \forall$ infinite n A_n intersects μx.

(2) This is just a concise restatement of (1) since A_n intersects μx iff $x \in st\, A_n$.

(3) By (2), since each $st\, A_n$ is closed. ○

The reason we're interested in $\lim_d \langle A_n \rangle$ is that ...

4.2 Proposition Every convergent sequence in $\mathcal{K}X$ converges to its limit set.

Proof:

Suppose $\langle A_n \rangle \to B$. Then for all infinite n, $A_n \in \mu B$ so $B = st\, A_n$; thus $B = \bigcap \{st\, A_n \mid n \text{ is infinite}\} = \lim_d \langle A_n \rangle$. ○

Bear in mind though that for a sequence $\langle A_n \rangle$ in $\mathcal{K}X$ $\lim_d \langle A_n \rangle$ need not belong to $\mathcal{K}X$ and even if it does, $\langle A_n \rangle$ need not converge to it with respect to h. As an example of the latter, if X is $\mathbb{R}$ and A_n is $\{0\}$ if n is even and $\{0,1\}$ if n is odd, then $\lim_d \langle A_n \rangle = \{0\}$, but of course $\langle A_n \rangle$ is not convergent. More can be said in the case of increasing or decreasing sequences however. In general ...

4.3 Proposition For a sequence $\langle A_n \rangle$ of nonempty subsets of X,

(1) $\bigcap_n \overline{A_n} \subseteq \lim_d \langle A_n \rangle$ with equality if $\langle A_n \rangle$ is decreasing.

(2) $\lim_d \langle A_n \rangle \subseteq \overline{\bigcup_n A_n}$ with equality if $\langle A_n \rangle$ is increasing.

Proof:

(1) '$\subseteq$' is trivial. Now further suppose $\langle A_n \rangle$ is decreasing. Then for $x \in \lim_d \langle A_n \rangle$, $x \in st\, A_m$ for infinite m; and hence for all *finite* n we have (as $A_m \subseteq A_n{}^*$) $x \in st\,(A_n{}^*) = \overline{A_n}$.

(2) '$\subseteq$' is trivial. And if $\langle A_n \rangle$ is increasing we easily see that $\bigcup_n A_n \subseteq \lim_d \langle A_n \rangle$, hence (as the latter is closed) $\overline{\bigcup_n A_n} \subseteq \lim_d \langle A_n \rangle$. ○

And in particular . . .

4.4 **Proposition** For a sequence $\langle A_n \rangle$ in $\mathcal{K}X$,

(1) If $\langle A_n \rangle$ is decreasing, $\langle A_n \rangle \to \bigcap_n A_n$.

(2) If $\langle A_n \rangle$ is increasing and bounded above in $\mathcal{K}X$, $\langle A_n \rangle \to \overline{\bigcup_n A_n}$.

Proof:
(1) Let $B = \bigcap_n A_n$. Taking any infinite m, we must show that $A_m \in \mu B$. Since $A_m \subseteq ns\, X^*$ (as $A_m \subseteq A_0{}^*$) it remains to show (recalling 2.1 of Chapter 2) that $st\, A_m = B$. For all finite n, $A_m \subseteq A_n{}^*$ so $st\, A_m \subseteq st\,(A_n{}^*) = A_n$; so $st\, A_m \subseteq B$. Conversely, $B^* \subseteq A_m$ so $B = st\, B^* \subseteq st\, A_m$.
(2) Let $B = \overline{\bigcup_n A_n}$ and let $C \in \mathcal{K}X$ with all $A_n \subseteq C$. Taking any infinite m, we must show $A_m \in \mu B$. $A_m \subseteq ns\, X^*$ as $A_m \subseteq C^*$, so it remains to show that $st\, A_m = B$. $A_m \subseteq B^*$ so $st\, A_m \subseteq st\, B^* = \overline{B} = B$. Conversely, for all finite n, $A_n{}^* \subseteq A_m$ so $A_n = st\,(A_n{}^*) \subseteq st\, A_m$; so $\bigcup_n A_n \subseteq st\, A_m$, so as the latter's closed, $B \subseteq st\, A_m$. ○

Also note that . . .

4.5 **Proposition** In $\mathcal{K}X$, if $\langle A_n \rangle \to B$ then $(\bigcup_n A_n) \cup B = \overline{\bigcup_n A_n} \in \mathcal{K}X$.

Proof:
Let $C = \bigcup_n A_n$. As $\langle A_n \rangle \to B$ then $\{A_n \mid n \in \omega\} \cup \{B\}$ is a compact subset of $\mathcal{K}X$ hence the union $C \cup B$ is a compact subset of X. We now show that it equals $\overline{C}$. By 4.3(2), $B \subseteq \overline{C}$ giving $C \cup B \subseteq \overline{C}$. For the reverse, if $x \in \overline{C} - C$ then where $c \in C^*$ with $c \in \mu x$, since $x \notin C$ it follows that c belongs to some A_n with n *infinite*, so $x \in st\, A_n = B$. ○

Another result of interest is . . .

4.6 **Proposition** For convergent sequences $\langle A_n \rangle$ and $\langle B_n \rangle$ in $\mathcal{K}X$,

$$(\forall n\ A_n \text{ intersects } B_n) \Rightarrow \lim \langle A_n \rangle \text{ intersects } \lim \langle B_n \rangle .$$

Proof:
Let $\langle A_n \rangle \to C$ and $\langle B_n \rangle \to D$. Assuming the left hand side, then taking infnite n, A_n and B_n intersect at some α; and then $st\, \alpha \in st\,(A_n) \cap st\,(B_n) = C \cap D$. ○

Finally we remark that most of this section can be generalized to the Vietoris topology ; in particular the proofs of the last three results were purely topological and the results therefore apply to arbitrary Hausdorff X, giving $\mathcal{K}X$ the Vietoris topology. In the purely topological setting one defines $x \in X$ to be a 'limit point' of $\langle A_n \rangle$ if for every neighbourhood N of x, A_n intersects N for all sufficiently large n. This is equivalent to asking that A_n intersect μx for all *infinite* n, i.e. to the condition that $x \in \bigcap \{st\, A_n \mid n \text{ is infinite}\}$, so the 'limit set' of $\langle A_n \rangle$ is thus $\bigcap \{st\, A_n \mid n \text{ is infinite}\}$ as before. Note also that if x has a *countable* neighbourhood basis (which is of course true in the *metric* case), x is a limit point of $\langle A_n \rangle$ iff there's a sequence $\langle a_n \rangle \to x$ with each $a_n \in A_n$.

5. The Map $C(X,Y) \to C(\mathcal{K}X, \mathcal{K}Y)$

Throughout let X and Y be metric spaces. In 8.1 of Chapter 1 we noted how each $f \in C(X,Y)$ induces a map $f_{\mathcal{K}} \in C(\mathcal{K}X, \mathcal{K}Y)$. In the next result we establish the continuity of this map $f \mapsto f_{\mathcal{K}}$ with respect to uniform topologies of the type described in 'Preliminaries' under 'Topologies of Uniform Convergence Over Subsets'. For the definition of the ideal $\mathcal{J}_{\mathcal{K}}$ induced by $\mathcal{J}$ below, consult Section 6 of Chapter 1.

5.1 Proposition For any ideal $\mathcal{J}$ on X, giving $C(X,Y)$ the $\mathcal{J}$-uniform topology and $C(\mathcal{K}X, \mathcal{K}Y)$ the $\mathcal{J}_{\mathcal{K}}$-uniform topology, ${}_{\mathcal{K}} : C(X,Y) \to C(\mathcal{K}X, \mathcal{K}Y)$ is continuous.

Proof:
Let $f \in C(X,Y)$ and take any $g \in \mu f$; so $\forall \alpha \in \nu \mathcal{J}\ \ g\,\alpha \approx f\,\alpha$. We must show that $g_{\mathcal{K}} \in \mu f_{\mathcal{K}}$, i.e. that $\forall A \in \nu\, \mathcal{J}_{\mathcal{K}}$, $g_{\mathcal{K}}\, A \approx f_{\mathcal{K}}\, A$. Since $A \in \nu\, \mathcal{J}_{\mathcal{K}}$ we know that $A \subseteq \nu \mathcal{J}$, hence $\forall \alpha \in A\ \ g\,\alpha \approx f\,\alpha$; hence clearly $g\,A \approx f A$, i.e. $g_{\mathcal{K}}\, A \approx f_{\mathcal{K}}\, A$. ○

Taking $\mathcal{J} = subCp\, X$ and using 7.1 of Chapter 1 (that $\mathcal{J}_{\mathcal{K}} = subCp\, \mathcal{K}X$), or taking $\mathcal{J} = Bd\, X$ and using 2.2 (that $\mathcal{J}_{\mathcal{K}} = Bd\, \mathcal{K}X$), we therefore have that . . .

5.2 Corollary The map ${}_{\mathcal{K}} : C(X,Y) \to C(\mathcal{K}X, \mathcal{K}Y)$ is continuous with respect to the compact-uniform topologies and the bounded-uniform topologies. ○

The next result shows that among the functional properties preserved by the map ${}_{\mathcal{K}} : C(X,Y) \to C(\mathcal{K}X, \mathcal{K}Y)$ is that of being Lipschitz of a given ratio ;

5.3 Proposition For $f \in C(X,Y)$, $r_{(f_{\mathcal{K}})} = r_f$.

Proof:
For all $A,B \in \mathcal{K}X$, each $a \in A$ is within distance $h(A,B)$ of some $b \in B$, hence $f\,a$ is within distance $r_f h(A,B)$ of $f\,b \in f\,B$; likewise vice versa so $h(f\,A, f\,B) \leq r_f h(A,B)$. So $r_{f_{\mathcal{K}}} \leq r_f$. And as X is embedded in $\mathcal{K}X$, $r_{f_{\mathcal{K}}} \geq r_f$. ○

Finally we'll note for use in the next section that . . .

5.4 Note Let $\mathcal{J}$ be an ideal on X and give $C(X,Y)$ the $\mathcal{J}$-uniform topology. Then if $\mathcal{J}$ is topological, the evaluation map $C(X,Y) \times X \to Y$ is continuous.

Proof:
If $(g,\alpha) \in \mu(f,x)$ then as $\alpha \in ns\, X^* \subseteq \nu \mathcal{J}$ and f is continuous, $g\,\alpha \approx f\,\alpha \approx f\,x$. ○

6. Union Functions

Throughout this section let Z be a metric space, $\mathcal{J}$ be an ideal on Z, and the set $F(Z,\mathcal{K}X)$ of functions $Z \to \mathcal{K}X$ have the $\mathcal{J}$-uniform topology, likewise the subspace $C(Z,\mathcal{K}X)$ of *continuous* functions. For $F \subseteq F(Z,\mathcal{K}X)$ and $z \in Z$ we define $Fz = \{fz \mid f \in F\}$, and with a similar proof to that of 9.1 of Chapter 1,

6.1 Note For $F \in \mathcal{K}F(Z,\mathcal{K}X)$ and $z \in Z$, $Fz \in \mathcal{K}\mathcal{K}X$ so $\bigcup(Fz) \in \mathcal{K}X$. ○

So for $F \in \mathcal{K}F(Z,\mathcal{K}X)$ we define $\bigcup F : Z \to \mathcal{K}X$ by $(\bigcup F)z = \bigcup(Fz) = \bigcup_{f \in F} fz$, which we'll call the <u>union function</u> of F. We thus have a map $\bigcup : \mathcal{K}F(Z,\mathcal{K}X) \to F(Z,\mathcal{K}X)$, and we find that . . .

6.2 Proposition The map $\bigcup : \mathcal{K}F(Z,\mathcal{K}X) \to F(Z,\mathcal{K}X)$ is continuous.

Proof:

Let $G \in \mu F$. We must show $\bigcup G \in \mu \bigcup F$, i.e. $\forall z \in \nu\mathcal{J}$ $(\bigcup G)z \approx (\bigcup F)^* z$, i.e. $\bigcup(Gz) \approx \bigcup(F^* z)$. By uniform continuity of $\bigcup : \mathcal{K}\mathcal{K}X \to \mathcal{K}X$ it suffices to show that $Gz \approx F^* z$ in $\mathcal{K}\mathcal{K}X$. And this holds since . . .

(1) $\forall g \in G$ $\exists f \in F$ with $g \in \mu f$, so $gz \approx fz \in F^* z$.

(2) $\forall h \in F^*$ $\exists f \in F$ with $h \in \mu f$, and $\exists g \in G$ with $g \in \mu f$, so $hz \approx fz \approx gz$ giving $hz \approx gz \in Gz$. ○

We'll say a subset S of $F(Z,\mathcal{K}X)$ is <u>closed under compact unions</u> if $\bigcup F \in S$ for all $F \in \mathcal{K}S$. In this case, by 6.2 the map $\bigcup : \mathcal{K}S \to S$ will be of course be continuous. If $\mathcal{J}$ is topological then $C(Z,\mathcal{K}X)$ is closed under compact unions ;

6.3 Proposition If $\mathcal{J}$ is topological then

(1) The map $\mathcal{K}C(Z,\mathcal{K}X) \times Z \to \mathcal{K}\mathcal{K}X$ in which $(F,z) \mapsto Fz$, is continuous.

(2) The map $\mathcal{K}C(Z,\mathcal{K}X) \times Z \to \mathcal{K}X$ in which $(F,z) \mapsto (\bigcup F)z$, is continuous.

(3) $C(Z,\mathcal{K}X)$ is closed under compact unions.

Proof:

(1) Suppose $(G,\alpha) \in \mu(F,z)$. Then $G\alpha \in \mu Fz$ since $\forall g \in G$ $\exists f \in F$ $g \in \mu f$ and $\forall f \in F$ $\exists g \in G$ $g \in \mu f$, and in general if $g \in \mu f$ then $g\alpha \in \mu fz$ by continuity of the evaluation map $C(Z,\mathcal{K}X) \times Z \to \mathcal{K}X$ (by 5.4).

(2) By (1) and continuity of the union map $\bigcup : \mathcal{K}\mathcal{K}X \to \mathcal{K}X$.

(3) For $F \in \mathcal{K}C(Z,\mathcal{K}X)$, $\bigcup F$ is continuous by (2). ○

Since $Bd\,Z$ is a topological ideal, and if Z is locally compact $subCp\,Z$ is a topological ideal, note that . . .

6.4 **Corollary** If $C(Z,\mathcal{K}X)$ has the bounded-uniform topology, or has the compact-uniform topology and Z is locally compact, $C(Z,\mathcal{K}X)$ is closed under compact unions and $\bigcup : \mathcal{K}C(Z,\mathcal{K}X) \to C(Z,\mathcal{K}X)$ is continuous. ○

Note that the part of 6.4 concerning the compact-uniform topology was proved independently in 9.3 of Chapter 1, since the compact-uniform topology here is the compact-open topology.

One subset of $F(Z,\mathcal{K}X)$ closed under compact unions *regardless* of the nature of $\mathcal{J}$ is the set $Lip_u(Z,\mathcal{K}X)$ of Lipschitz maps $f : Z \to \mathcal{K}X$ with ratio $r_f \leq u$, for $u \in [0,\infty)$;

6.5 **Proposition** For $F \in \mathcal{K}F(Z,\mathcal{K}X)$, $r_{\bigcup F} \leq \bigvee_{f\in F} r_f$.

So for any $u \in [0,\infty)$, $Lip_u(Z,\mathcal{K}X)$ is closed under compact unions.

Proof:

For all $y,z \in Z$,

$$\begin{aligned} h(\bigcup F y, \bigcup F z) &= h(\bigcup_{f\in F} fy, \bigcup_{f\in F} fz) \\ &\leq \bigvee_{f\in F} h(fy,fz) \qquad \text{(by 1.6)} \\ &\leq \bigvee_{f\in F} r_f h(y,z) \\ &= (\bigvee_{f\in F} r_f)\, h(y,z) . \end{aligned}$$

So $r_{\bigcup F} \leq \bigvee_{f\in F} r_f$. ○

Using the above results together with those of Section 5 we now consider the matter of a nonempty compact set F of continuous operations on X inducing an operation on $\mathcal{K}X$. For the rest of the section let $\mathcal{I}$ be an ideal on X, and give $C(X,X)$ the $\mathcal{I}$-uniform topology and $C(\mathcal{K}X,\mathcal{K}X)$ the $\mathcal{I}_{\mathcal{K}}$-uniform topology. Now by 5.1 the map $_{\mathcal{K}} : C(X,X) \to C(\mathcal{K}X,\mathcal{K}X)$ is continuous, so it maps each $F \in \mathcal{K}C(X,X)$ to $F_{\mathcal{K}} = \{f_{\mathcal{K}} \mid f \in F\} \in \mathcal{K}C(\mathcal{K}X,\mathcal{K}X) \subseteq \mathcal{K}F(\mathcal{K}X,\mathcal{K}X)$, from which in turn we get (considering $Z = \mathcal{K}X$ and $\mathcal{J} = \mathcal{I}_{\mathcal{K}}$) the union function $\bigcup(F_{\mathcal{K}}) \in F(\mathcal{K}X,\mathcal{K}X)$, which we'll naturally call the <u>union function</u> of F, denoted by $\bigcup F$ and given by $(\bigcup F)A = \bigcup(F_{\mathcal{K}} A) = \bigcup_{f\in F} fA = FA$ where $FA = \{fa \mid f \in F \text{ and } a \in A\}$. In this way then, each nonempty compact set F of continuous operations on X induces an operation $\bigcup F$ on $\mathcal{K}X$. As the following shows, this map $F \mapsto \bigcup F$ is continuous, and as long as $\mathcal{I}$ is topological $\bigcup F$ is continuous ;

6.6 **Proposition**

(1) The map $\bigcup : \mathcal{K}C(X,X) \to F(\mathcal{K}X,\mathcal{K}X)$ is continuous.

(2) If $\mathcal{I}$ is topological, $\bigcup F$ is continuous for all $F \in \mathcal{K}C(X,X)$, hence we have a continuous map $\bigcup : \mathcal{K}C(X,X) \to C(\mathcal{K}X,\mathcal{K}X)$.

Proof:
(1) It's the composition of the map $_{\mathscr{K}} : \mathscr{K}C(X,X) \to \mathscr{K}C(\mathscr{K}X, \mathscr{K}X)$ followed by the map $\bigcup : \mathscr{K}F(\mathscr{K}X, \mathscr{K}X) \to F(\mathscr{K}X, \mathscr{K}X)$; and the latter is continuous by 6.2 (putting $Z = \mathscr{K}X$ and $\mathscr{J} = \mathscr{I}_{\mathscr{K}}$), whilst the former is continuous as it's the map induced by the continuous map $_{\mathscr{K}} : C(X,X) \to C(\mathscr{K}X, \mathscr{K}X)$ (apply 8.1 of Chapter 1 with X and Y replaced there by $C(X,X)$ and $\mathscr{K}C(\mathscr{K}X, \mathscr{K}X)$).
(2) $\bigcup F$ is $\bigcup(F_{\mathscr{K}})$, which is continuous by (3) of 6.3 (putting $Z = \mathscr{K}X$ and $\mathscr{J} = \mathscr{I}_{\mathscr{K}}$ (which is topological as $\mathscr{I}$ is)). ○

In particular (2) applies if $\mathscr{I} = Bd\,X$ (giving the bounded-uniform topologies on $C(X,X)$ and $C(\mathscr{K}X, \mathscr{K}X)$), or if $\mathscr{I} = subCp\,X$ (giving the compact-uniform topologies) and X is locally compact (so that $\mathscr{I}$ is topological). The continuity of $\bigcup : \mathscr{K}C(X,X) \to C(\mathscr{K}X, \mathscr{K}X)$ in the latter case was proved independently in 9.4 of Chapter 1, since compact-uniform equates with compact-open.

Lastly we have the following results on Lipschitz ratios and maps ;

<u>6.7</u> **<u>Proposition</u>** For $F \in \mathscr{K}C(X,X)$, $r_{\bigcup F} \leq \bigvee_{f \in F} r_f$. So for any $u \in [0,\infty)$, we have a continuous map $\bigcup : \mathscr{K}Lip_u(X,X) \to Lip_u(\mathscr{K}X, \mathscr{K}X)$.

Proof:
The first statement holds by 6.5 and 5.3 ($r_{\bigcup F} = r_{\bigcup(F_{\mathscr{K}})} \leq \bigvee_{f \in F} r_{(f_{\mathscr{K}})} = \bigvee_{f \in F} r_f$) and the second follows, continuity being provided by (1) of 6.6 . ○

<u>6.8</u> **<u>Proposition</u>** For $F \in \mathscr{K}C(X,X)$, $r_{\bigcup F}$ is the least $r \in [0,\infty]$ such that

$$\forall x,y \in X\ \forall f \in F\ \exists g \in F\ \ d(f\,x, g\,y) \leq r\, d(x,y).$$

Proof:
In other words we must show that for $r \in [0,\infty]$,

$$r_{\bigcup F} \leq r \Leftrightarrow \forall x,y \in X\ \forall f \in F\ \exists g \in F\ \ d(f\,x, g\,y) \leq r\, d(x,y).$$

$\Rightarrow$: Let $x,y \in X$ and $f \in F$. Then $h(F\{x\}, F\{y\}) \leq r\, h(\{x\},\{y\}) = r\, d(x,y)$, so as $f\,x \in F\{x\}$ there's $g \in F$ with $d(f\,x, g\,y) \leq r\, d(x,y)$.
$\Leftarrow$: Let $A,B \in \mathscr{K}X$. We must show $h(FA,FB) \leq r\, h(A,B)$. For all $f \in F$ and $a \in A$, there's $b \in B$ with $d(a,b) \leq h(A,B)$ and there's $g \in F$ with $d(f\,a, g\,b) \leq r\, d(a,b)$; so we have $g\,b \in FB$ with $d(f\,a, g\,b) \leq r\, h(A,B)$. So, every element of FA is within distance $r\, h(A,B)$ of an element of FB . Likewise vice versa, as required. ○

As a corollary, for $F \in \mathscr{K}C(X,X)$ $\bigcup F$ is a contraction iff for some $r < 1$, $\forall x,y \in X\ \forall f \in F\ \exists g \in F\ \ d(f\,x, g\,y) \leq r\, d(x,y)$.

Chapter 3

Hutchinson's Invariant Sets

0. Introduction

Perhaps the most primitive notion of a 'self-similar' set is that of a nonempty compact subset K of a metric space such that K is the union of a finite number of smaller images of itself, say $K = \bigcup_{f \in F} fK$ for some finite set F of contractions of the space. In the 1981 paper [Hu], Hutchinson developed the theory thereof, in particular proving the existence of a *unique* such K for any nonempty finite set F of contractions of a complete metric space by a beautifully simple application of Banach's contraction mapping theorem. In Hutchinson's terminology the set K is said to be 'invariant' with respect to F. In this chapter we give a treatment of the basic theory generalized to nonempty *compact* sets F of contractions, making various other contributions along the way, among them a further generalization from contractions to 'reductions'.

Throughout, X will denote a nonempty complete metric space with metric d, and $\mathcal{K}X$ will as usual denote the set of nonempty compact subsets of X, equipped with the Hausdorff metric h, with which familiarity in the form of Chapter 2 will be assumed. The set $C(X,X)$ of continuous maps $X \to X$ is given the compact-uniform topology. Recall from 'Preliminaries' that the Lipschitz ratio of $f: X \to X$ is denoted by r_f, and that on the set Lip_1X of Lipschitz maps with ratio ≤ 1, the compact-uniform topology coincides with the topology of pointwise convergence, making Lip_1X a topological monoid in which, for $f \in Lip_1X$, $\mu f = \{g \in Lip_1X^* \mid \forall x \in X \ \forall \alpha \in \mu x \quad g\,\alpha \in \mu\, f x\}$.

A summary of the chapter follows. In Section 1, using previous work on 'union functions' the existence of a unique invariant set K is established for each 'admissible' set of contractions F. In addition we consider the benefit of having extended the theory to admissible *compact* sets of contractions rather than just keeping to *finite* sets , and give one or two basic results. Section 2 concerns the way in which not only is K composed of the smaller images fK ($f \in F$) of itself (its so-called 'first-level images'), but in turn these images are each composed of still *smaller* images of K (its 'second-level images'), and so on ad infinitum. In Section 3 we see how each sequence in F determines a point of K and consider the resulting 'code map' $k: F^{\omega} \to K$. Section 4 shows how the 'periodic points' (namely the fixed points of compositions of elements of F) are dense in K. Section 5 concerns the connectedness of K and describes the characterization due to Hata in [Ha], along with various other results. In Section 6 we show that as long as F satisfies the 'homeomorphism condition' (which in particular holds if all the elements of F are homeomorphisms of X) then K is either regular or residual (i.e. has empty interior). One sufficient condition for residuality of K is given, a more widely applicable condition appearing in Section 7, which is concerned with the especially tidy situation in which the first-level images of K are, within the space K, regular and nonoverlapping, thus forming a 'tiling' of K. Section 8 deals with the continuity of the 'attractor map' which sends each admissible set F of contractions to its invariant set (or 'attractor') K_F. In Section 9 we show that practically all of the preceding theory generalises from contractions to 'reductions', a weaker class of maps which nevertheless have some of the properties of contractions, in particular obeying a fixed point theorem which generalises Banach's contraction mapping theorem. Finally, Section 10 comprises an assortment of notes, questions, and suggestions for further work.

1. Existence of Invariant Sets

Recall from Section 6 of Chapter 2 that the 'union function' of $F \in \mathscr{K}C(X,X)$ is the map $\bigcup F : \mathscr{K}X \to \mathscr{K}X$ given by $(\bigcup F)A = \bigcup_{f \in F} fA = \{fa \mid f \in F \text{ and } a \in A\} = FA$ (and note that $(\bigcup F)^n A$ may be written as $F^n A$). Hutchinson showed in [Hu] that the union function of every nonempty finite set of *contractions* of X is a contraction of $\mathscr{K}X$ and therefore has a unique fixed point by Banach's contraction mapping theorem, $\mathscr{K}X$ being complete as X is. We can extend this to certain nonempty *compact* sets of contractions of X as follows. Namely, calling a set F of contractions of X **admissible** if it is nonempty and compact and the Lipschitz ratios of its elements have a common upper bound less than 1, i.e. $\bigvee_{f \in F} r_f < 1$ (which in particular holds if F is finite), then since by 6.7 of Chapter 2 $r_{\bigcup F} \leq \bigvee_{f \in F} r_f$, we have $r_{\bigcup F} < 1$, so . . .

1.1 Proposition The union function of any admissible set F of contractions of X is a contraction of $\mathscr{K}X$, hence there exists a unique $K \in \mathscr{K}X$ such that $FK = K$. Furthermore, for all $A \in \mathscr{K}X$, $\langle F^n A \rangle \to K$. ○

The set K above is in other words the unique fixed point of $\bigcup F$, and we may denote it more fully by K_F, though with F in mind we'll usually just write K. K is called the **attractor** of F, or in Hutchinson's terminology is said to be **invariant** with respect to F. It's tempting to say K is 'self-similar' with respect to F in so far as it's composed of the 'smaller images' fK of itself, but that can be a little too suggestive. After all, these images needn't *look* like K at all (we needn't be dealing with similitudes of $\mathbb{R}^n$) and they *may* overlap and interfere with one another, resulting in visual confusion. Hutchinson himself uses the terminology 'self-similar with respect to F' for something stronger, namely requiring not only that K be invariant with respect to F but also that, where δ is the Hausdorff dimension of K and H^δ is Hausdorff δ-dimensional measure, then with respect to H^δ, K must be non-null whilst for distinct $f,g \in F$, $fK \cap gK$ must be null ; roughly speaking then, K must be of substance and the images fK of K must be essentially disjoint. Hutchinson points out that even this is still quite weak. So all in all, we play safe by keeping to the modest term 'invariant'. In fact this term is all the more agreeable and descriptive if considered as coming from its usage in topological dynamics. One can think of F inducing a discrete 'multiple semiflow' in X, thinking of each $x \in X$ becoming (over each time step) its images fx ($f \in F$), and one naturally says $A \subseteq X$ is 'invariant' with respect to F if it remains constant, i.e. $\bigcup_{f \in F} fA = A$, i.e. $FA = A$. In this sense our K is the unique invariant element of $\mathscr{K}X$ then. The semiflow picture also suggests what can be thought of as the *continuous* analogue of Hutchinson's theory in which the semiflows are no longer *discrete* but continuous; see note 10.17 for more detail.

Having introduced invariant sets, the following four pages comprise twenty-four examples in $\mathbb{R}^2$, which may be useful in visualizing later theory. Produced on a Sun workstation using the random iteration algorithm (see [Ba], or [Wic]), most of these have not appeared in the literature before. In each case the number of contractions involved is stated, and in most cases the reader should have little trouble in qualitatively *determining* the contractions by identifying the images fK of K for $f \in F$. Except for some of the affine maps used in Figures 15, 16 and 17, all the maps involved are direct similitudes (see Appendix 9). Figures 1 to 16 show examples having the rather special property of being *tiled* by their images fK ($f \in F$), a property studied in Section 7. They are also all residual, and seven of them are totally disconnected. Figures 9 and 10 show examples from a family described in [Ba] where many other pictures can be found, including one similar to Figure 3. Figures 7, 8 and 14 show spiral-based invariant sets of a type described in the recreational article [Wic] where further examples may be found. Figure 17 shows a fern (examples of which originated with Barnsley) which is *not* tiled by its images fK; there is overlap; likewise in Figure 18 which shows another spiral-based set. So far, all the sets involved have been 'designed', and are consequently rather exceptional. More *typical* examples are shown in Figures 19 to 24 where the contractions were chosen at *random* (subject to certain restrictions), the results being typically somewhat chaotic.

The set of admissible sets of contractions of X will be denoted by ***Admis*** X, and we point out right away that *every* nonempty compact set can be trivially realized as the attractor of some admissible F as follows. For $x \in X$ let c_x denote the constant map $X \to X$ with value x. Then . . .

<u>1.2</u> <u>Note</u> For $A \in \mathcal{K}X$, $A = K_F$ for $F = \{c_a \mid a \in A\}$, since $\bigcup F$ is the constant map with value A.

Proof:

F *is* an admissible set of contractions, being compact as it's the image of compact A under the embedding $x \mapsto c_x$ of X in $C(X,X)$. And $\bigcup F$ is clearly constant with value A, so has fixed point A. ∘

A trivial consequence is that the sets of the form K_F for *finite* F are dense in $\mathcal{K}X$, simply because they include all the nonempty finite sets, and these are dense in $\mathcal{K}X$.

One benefit of having generalized from finite sets F of contractions to admissible *compact* sets of contractions is that Barnsley's idea of an 'IFS with condensation' (see [Ba], page 93) can be coped for in the following sense. Barnsley essentially states that if F is a nonempty finite set of contractions of X and $A \in \mathcal{K}X$ then we get a contraction of $\mathcal{K}X$ by $B \mapsto (FB) \cup A$, hence this map has a unique fixed point. The fact that the map in question is a contraction of $\mathcal{K}X$ can be seen on the one hand from its being the

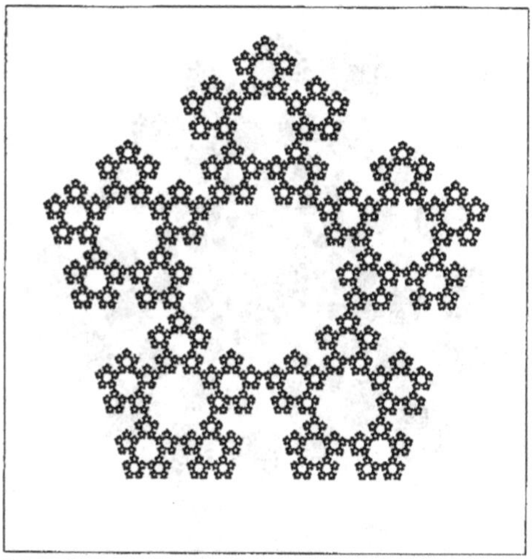

Fig. 1 : 5 contractions

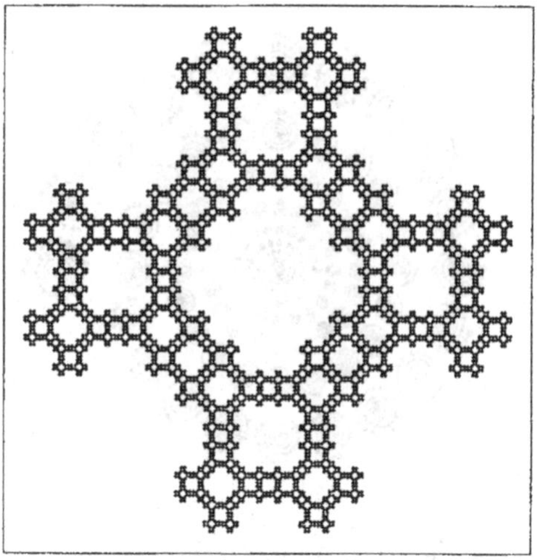

Fig. 2 : 4 contractions

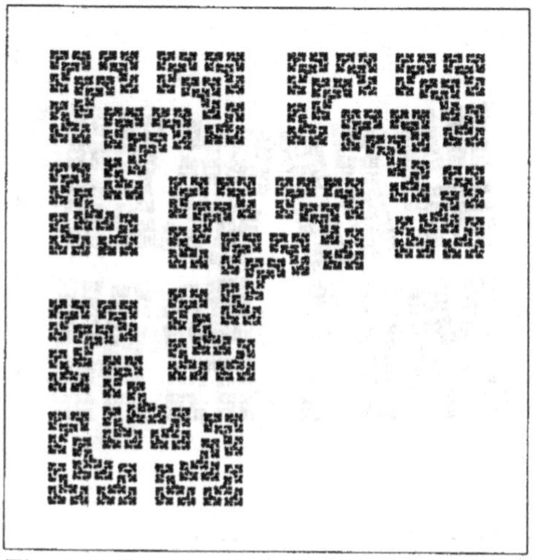

Fig. 3 : 4 contractions

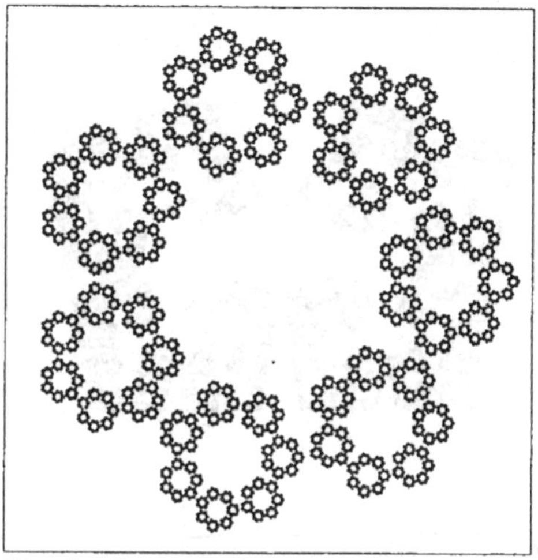

Fig. 4 : 7 contractions

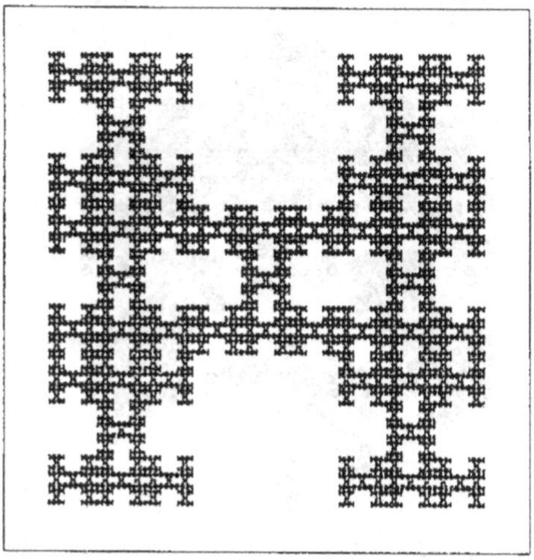

Fig. 5 : 7 contractions

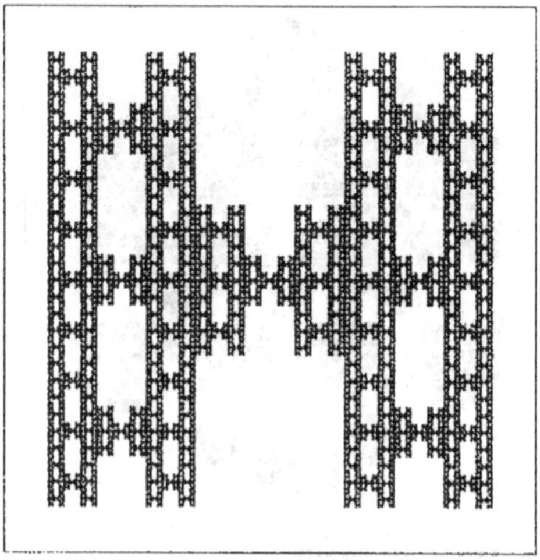

Fig. 6 : 7 contractions

Fig. 7 : 2 contractions

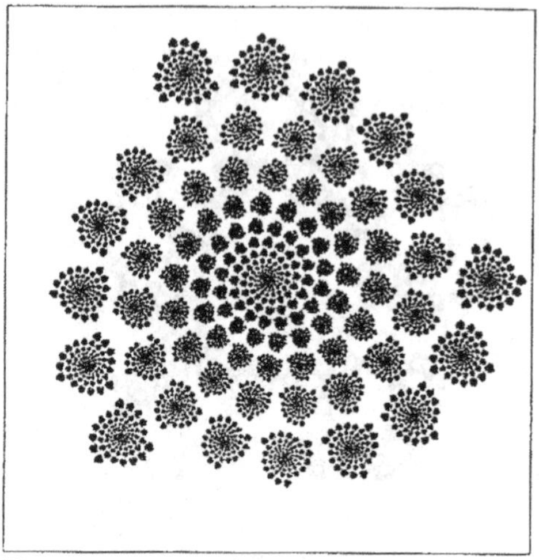

Fig. 8 : 2 contractions

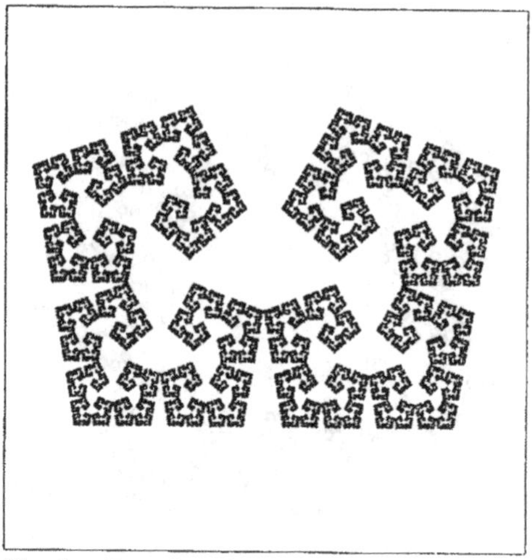

Fig. 9 : 2 contractions

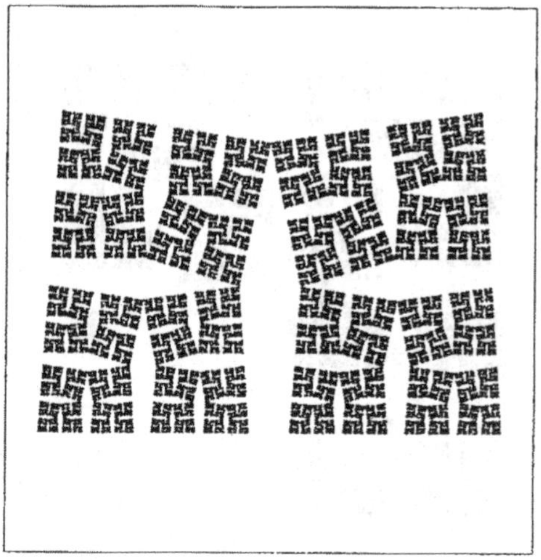

Fig. 10 : 2 contractions

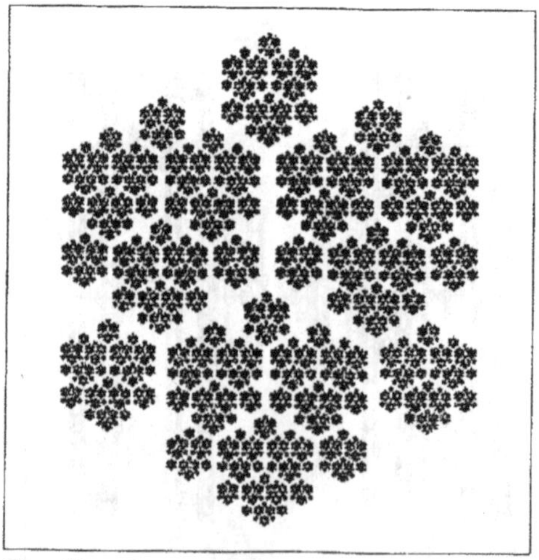

Fig. 11 : 6 contractions

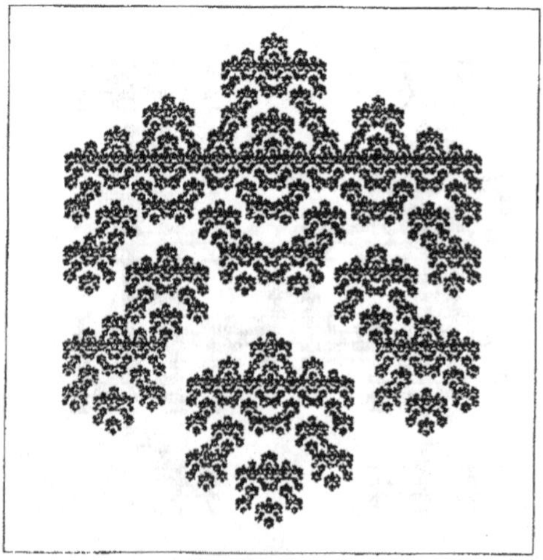

Fig. 12 : 6 contractions

Fig. 13 : 4 contractions

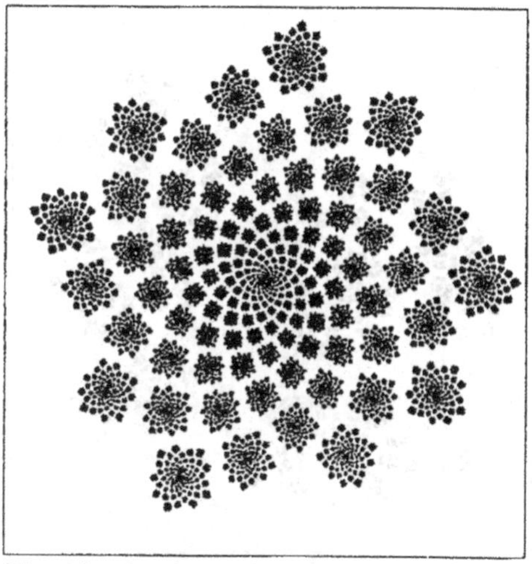

Fig. 14 : 2 contractions

Fig. 15 : 4 contractions

Fig. 16 : 4 contractions

Fig. 17 : 4 contractions

Fig. 18 : 2 contractions

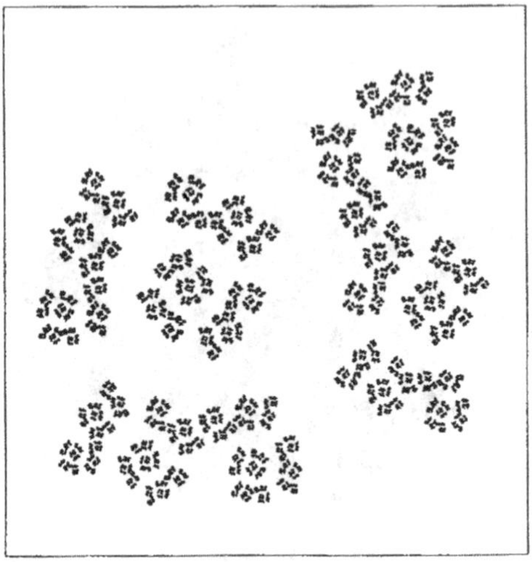

Fig. 19 : 3 contractions

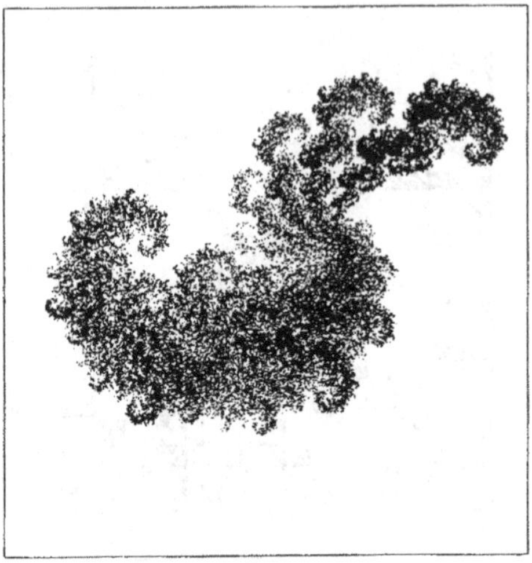

Fig. 20 : 4 contractions

Fig. 21 : 4 contractions

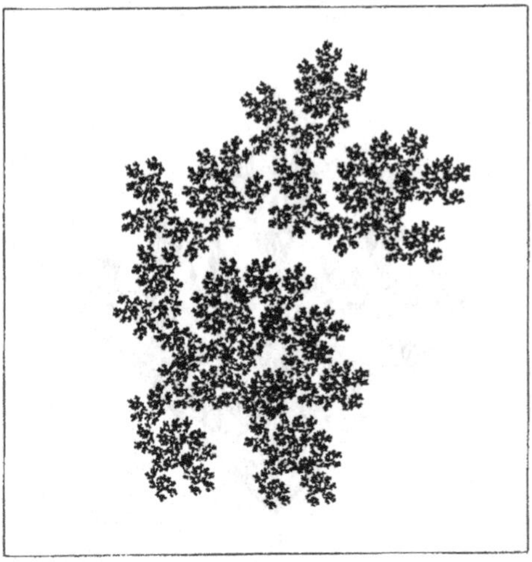

Fig. 22 : 4 contractions

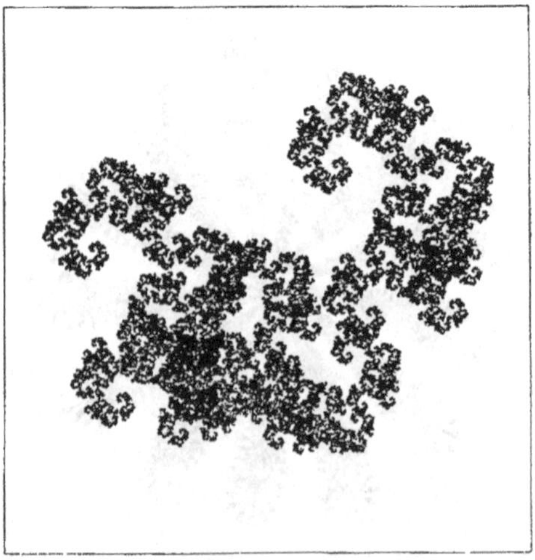

Fig. 23 : 4 contractions

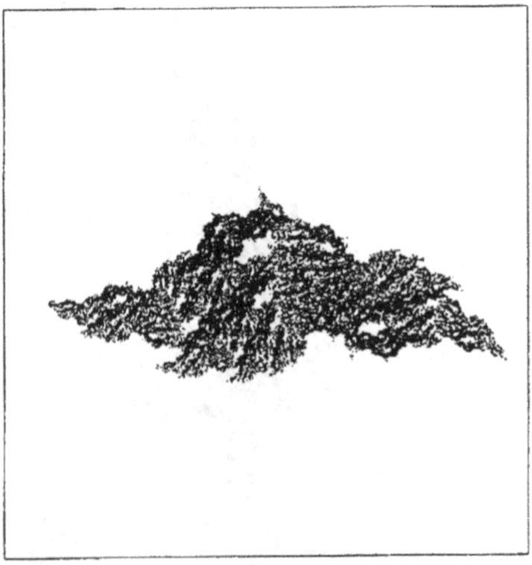

Fig. 24 : 4 contractions

union function of $\{\bigcup F, c_A\}$ with both maps $\bigcup F$ and c_A (the constant map with value A) being contractions of $\mathscr{K}X$, and using 6.5 of Chapter 2. However, by breaking down the contraction c_A of $\mathscr{K}X$ into the contractions c_a of X where $a \in A$, the map above is actually the contraction induced by $F \cup \{c_a \mid a \in A\} \in Admis\ X$. This is given in the next proposition along with a description of the invariant set. Some useful terminology we'll introduce beforehand is that the ***F*-range** of $A \in \mathscr{K}X$ is the closure of A under the elements of F and is denoted by A^F. Note that $A^F = \bigcup_n F^n A$, the union of the iterates $F^n A = (\bigcup F)^n A$ of A under the contraction $\bigcup F$ of $\mathscr{K}X$.

1.3 Proposition For $F \in Admis\ X$ and $A \in \mathscr{K}X$, where $G = \{c_a \mid a \in A\}$ we have

(1) $F \cup G \in Admis\ X$, with $\forall B \in \mathscr{K}X\ (F \cup G)B = (FB) \cup A$.

(2) $K_{F \cup G} = A^F \cup K_F = \overline{A^F}$ = the smallest closed expansion of A closed under all elements of F.

Proof:

(1) F and G are compact hence so is $F \cup G$. And $\bigvee_{f \in F \cup G} r_f = \bigvee_{f \in F} r_f$. So $F \cup G$ is admissible. And $\forall B \in \mathscr{K}X$, $(F \cup G)B = FB \cup GB = FB \cup A$.

(2) $\langle F^n A\rangle \to K_F$ so by 4.5 of Chapter 2 $(\bigcup_n F^n A) \cup K_F = \overline{\bigcup_n F^n A} \in \mathscr{K}X$, i.e. $A^F \cup K_F = \overline{A^F} \in \mathscr{K}X$. $\overline{A^F}$ is closed under all elements of F (since A^F is and the elements of F are continuous) and is hence clearly the *smallest* closed expansion of A closed under all $f \in F$. Lastly, $K_{F \cup G} = A^F \cup K_F$ as $(F \cup G)(A^F \cup K_F) = F(A^F \cup K_F) \cup A = F(A^F) \cup FK_F \cup A = A \cup F(A^F) \cup K_F = A^F \cup K_F$ noting that $A \cup F(A^F) = A^F$. ○

Changing the emphasis slightly, the above in particular gives the description $\overline{A^F} = A^F \cup K_F$ of $\overline{A^F}$, the closure of the union of A's orbit under $\bigcup F$, and we can use this in proving the following alternative characterization of K_F;

1.4 Proposition For $F \in Admis\ X$, K_F is the smallest nonempty closed set which is closed under all elements of F.

Proof:

We know that K_F is a nonempty closed set closed under all elements of F. And if C is too, then letting A be a nonempty compact subset of C (for example a singleton) we have $K_F \subseteq A^F \cup K_F = \overline{A^F} \subseteq \overline{C} = C$ so $K_F \subseteq C$. ○

It easily follows that . . .

1.5 Note For $F, G \in Admis\ X$, $F \subseteq G \Rightarrow K_F \subseteq K_G$. ○

Unless otherwise implied, assume for the rest of the chapter that F is an admissible set of contractions of X with attractor K, and $r = \bigvee_{f \in F} r_f$. Since the literature has largely only dealt with *finite* F (though [Ha] also considers *countable* F), credits to various results implicitly refer to this case. The generalizations are usually quite straightforward.

The next proposition gives two results we'll be using frequently later on, particularly the first. For $A \in \mathscr{K}X$ we say F **reduces** A if $FA \subseteq A$, i.e. if A is closed under all $f \in F$. Since the action of F preserves $\subseteq$ (i.e. $A \subseteq B \Rightarrow FA \subseteq FB$), note that F reduces A iff $\langle F^n A \rangle$ is decreasing. We say F **expands** A if $A \subseteq FA$, equivalently if $\langle F^n A \rangle$ is increasing.

1.6 Proposition (Monotonic approximation of K from above and below.) For $A \in \mathscr{K}X$,

(1) If F reduces A then $K = \bigcap_n F^n A$, and in particular $K \subseteq A$.

(2) If F expands A then $K = \overline{\bigcup_n F^n A} = \overline{A}^F$, and in particular $A \subseteq K$.

Proof:
(1) $\langle F^n A \rangle$ is decreasing, so by 4.4(1) of Chapter 2 its limit K is $\bigcap_n F^n A$.

(2) $\langle F^n A \rangle$ is increasing, so by 4.4(2) of Chapter 2 its limit K is $\overline{\bigcup_n F^n A}$. ○

As an example of how these results can be used in practice, we consider the 'Sierpinski Gasket' (more correctly, *a* Sierpinski Gasket), perhaps the most widely appearing example of an invariant set in the literature. In this example, $X = \mathbb{R}^2$ and $F = \{f_1, f_2, f_3\}$ where the maps f_i respectively contract about the vertices v_i of an equilateral triangle by factor 1/2. Letting A denote the (solid) triangle involved, A is reduced by F, hence $K = \bigcap_n F^n A$. The first few iterates $F^n A$ of A are shown below, converging to the limit K. Note that the application of $\bigcup F$ at each stage has the effect of removing the 'middle' open triangle of each triangle remaining. K is thus what remains after $\bigcup F$ has done its worst ;

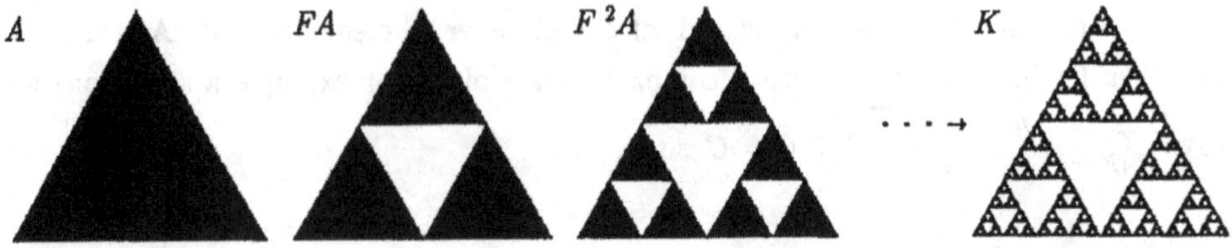

On the other hand, F *expands* the boundary ∂A of A and therefore $K = \overline{\bigcup_n F^n \partial A}$.

The first few iterates of ∂A are shown below, along with the limit K again. This time the application of $\bigcup F$ at each stage has the effect of adding in middle triangle boundaries. It should be emphasized that K is not the *union* of the iterates but the *closure* of the union ;

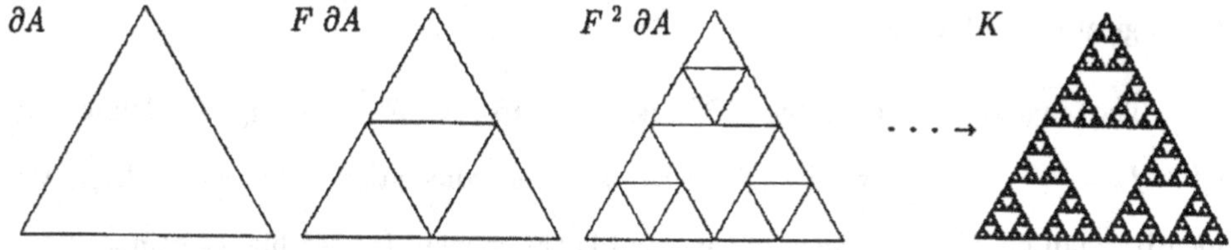

The last propostion can be generalized to the following, though it is used less frequently in practice ;

<u>1.7</u> **<u>Proposition</u>** (Approximation of K from above and below.) For $A \in \mathcal{K}X$,

(1) If $K \subseteq A$ then $K = \bigcap_n F^n A$.

(2) If $A \subseteq K$ then $K = \overline{\bigcup_n F^n A} = \overline{A^F}$.

Proof:
(1) Since in general $B \subseteq C \Rightarrow FB \subseteq FC$, and $FK = K$, we get $\forall n \;\; K \subseteq F^n A$, so $K \subseteq \bigcap_n F^n A$. And by 4.3(1) of Chapter 2, $\bigcap_n F^n A \subseteq \lim \langle F^n A \rangle = K$.

(2) As $A \subseteq K$ with K closed under the elements of F, $A^F \subseteq K$, so $\overline{A^F} \subseteq K$. And by 4.3(2) of Chapter 2, $K = \lim \langle F^n A \rangle \subseteq \overline{\bigcup_n F^n A} = \overline{A^F}$. ○

Since K is itself reduced and expanded by F, the above gives that K is the smallest nonempty compact set *reduced* by F, and also the largest nonempty compact set *expanded* by F. The first of these facts follows also from 1.4. Lastly, referring to the F-range of $\{x\}$ as the **F-range of x** and denoting it just by x^F, note by 1.7(2) with $A = \{x\}$ that . . .

<u>1.8</u> **<u>Corollary</u>** For $x \in K$, K is the closure of the F-range of x. ○

2. nth-Level Images

As described in Appendix 2, since F forms a Hausdorff space then the set $F^{<\omega}$ of finite sequences in F naturally forms a Hausdorff monoid under concatenation and a topology in which each set of the form F^n is open and has the product topology. In the following we'll assume familiarity with the notation concerning finite sequences as described in the first paragraph of Appendix 2, and also familiarity with Appendix 1 on topological monoid actions.

The composition map on $F^{<\omega}$ is the map $\circ : F^{<\omega} \to Lip_1X$ defined by $\circ f = f_0 \circ \cdots \circ f_{n-1}$ where $|f| = n$, with the convention that $\circ\emptyset = Id_X$, the identity map on X. The range of this map is the monoid (under binary composition) generated by F, denoted by ***Monoid F***, the structural aspect being that . . .

2.1 Note The composition map $\circ : F^{<\omega} \to Lip_1X$ is the unique topological monoid action of $F^{<\omega}$ on X such that the action of each $\langle f\rangle$ (for $f \in F$) is f.

Proof:
$\circ\emptyset = Id_X$ and clearly $\circ(fg) = (\circ f)\circ(\circ g)$, so $\circ$ is a monoid action of $F^{<\omega}$ on X, also clearly the *unique* one such that the action of each $\langle f\rangle$ (for $f \in F$) is f. And the action is topological since if $g \in \mu f$ and $\alpha \in \mu x$, then using continuity of binary composition in Lip_1X, we have $\circ g \in \mu(\circ f)$ and hence $(\circ g)\alpha \in \mu(\circ f)x$ as required. ○

For $f \in F^{<\omega}$ and $x \in X$ we'll write $(\circ f)x$ as fx. For $A \in \mathscr{K}X$ we also define $fA = (\circ f)A = \{fa \mid a \in A\}$. As described in Appendix 1 (namely in A1.2(2)), this gives the naturally induced monoid action of $F^{<\omega}$ on $\mathscr{K}X$. In more detail,

2.2 Corollary The monoid $F^{<\omega}$ topologically acts on $\mathscr{K}X$ by $(f,A) \mapsto fA$, this being the unique action such that the action of each $\langle f\rangle$ (for $f \in F$) is $f_{\mathscr{K}}$. ○

Be careful to note that in $\circ f$, the functions are applied in the *reverse* order to that in which they appear in f. Also note that each non-trivial composition $\circ f$ (i.e. $f \neq \emptyset$) is a contraction of X with ratio $\leq r^{|f|}$ so in general we have

$$h(fA,fB) \leq r^{|f|}h(A,B) \quad \text{and} \quad \operatorname{diam} fA \leq r^{|f|}\operatorname{diam} A .$$

Each such $\circ f$ also has a unique fixed point, and in Section 4 we'll find that these fixed points are dense in K, a result due to Hutchinson.

Now to the main definition of this section. For $A \in \mathscr{K}X$ and $n \in \omega$, an nth-level image of A (with respect to F) is a set of the form fA where $f \in F^n$; and a set of this form for some n is called a finite-level image of A. It's easily seen that . . .

2.3 Proposition For all $A \in \mathscr{K}X$ and $n \in \omega$, F^nA is the union of the nth-level images of A; $F^nA = \bigcup_{f \in F^n} fA$. ○

The above can be seen for example as follows. *Monoid F* acts on X in the natural way, hence (as described in Appendix 1) $\mathscr{K}$*Monoid F* forms a topological monoid under $F \circ G = \{f \circ g \mid f \in F \text{ and } g \in G\}$ and acts on $\mathscr{K}X$ by $(F,A) \mapsto FA$. So in general $G_1 \cdot\cdot\cdot G_n A = (G_1 \circ \cdot\cdot\cdot \circ G_n)A = \bigcup\{(g_1 \circ \cdot\cdot\cdot \circ g_n)A \mid \text{all } g_i \in G_i\}$, which putting all $G_i = F$ gives 2.3. And as $F^nK = K$ we have the basic result that . .

2.4 Corollary [Hu] For all $n \in \omega$, K is the union of its nth-level images. ○

The set of nth-level images of a set can be obtained in the following interesting way. Since F is an admissible set of contractions of X, it follows using $r_{(f_{\mathscr{K}})} = r_f$ and continuity of $_{\mathscr{K}} : C(X,X) \to C(\mathscr{K}X, \mathscr{K}X)$ that $F_{\mathscr{K}} = \{f_{\mathscr{K}} \mid f \in F\}$ is an admissible set of contractions of $\mathscr{K}X$, therefore giving a contraction $\bigcup(F_{\mathscr{K}})$ of $\mathscr{K}\mathscr{K}X$ by $(\bigcup(F_{\mathscr{K}}))\mathscr{A} = F_{\mathscr{K}}\mathscr{A} = \{g A \mid g \in F_{\mathscr{K}} \text{ and } A \in \mathscr{A}\} = \{fA \mid f \in F \text{ and } A \in \mathscr{A}\}$. For $A \in \mathscr{K}X$, the set $\{g A \mid g \in F^n\}$ of nth-level images of A is in $\mathscr{K}\mathscr{K}X$, being the image of nonempty compact F^n under the continuous map $g \mapsto g A$, and we find that $F_{\mathscr{K}}\{g A \mid g \in F^n\} = \{fg A \mid f \in F \text{ and } g \in F^n\} = \{g A \mid g \in F^{n+1}\}$ = the set of $(n+1)$th-level images of A. By induction starting with the set $\{A\}$ of 0th-level images of A, the set of nth-level images of A is therefore $F_{\mathscr{K}}{}^n\{A\}$. Also note that the invariant set of $F_{\mathscr{K}}$ is $\{\{x\} \mid x \in K\}$ since $F_{\mathscr{K}}\{\{x\} \mid x \in K\} = \{f\{x\} \mid f \in F \text{ and } x \in K\} = \{\{fx\} \mid f \in F \text{ and } x \in K\} = \{\{y\} \mid y \in FK = K\}$; this could have been anticipated in view of the fact that the actions of the elements of F on X are embedded in the actions of the elements of $F_{\mathscr{K}}$ by $x \mapsto \{x\}$ and $f \mapsto f_{\mathscr{K}}$ (in the sense that the former embeds the metric space X in $\mathscr{K}X$, and $f_{\mathscr{K}}\{x\} = \{fx\}$). One particular consequence of all this is therefore that . . .

2.5 Proposition For $A \in \mathscr{K}X$, $\langle\{g A \mid g \in F^n\}\rangle \to \{\{x\} \mid x \in K\}$ in $\mathscr{K}\mathscr{K}X$. ○

The following is a simple observation on how the $(n+1)$th-level images $fg\,A$ of A relate to the nth-level image fA of A when A is reduced or expanded by F;

2.6 Note For $A \in \mathscr{K}X$ and $f \in F^{<\omega}$,

(1) If F reduces A, $fA \supseteq \bigcup_{g \in F} fg\,A$.

(2) If F expands A, $fA \subseteq \bigcup_{g \in F} fg\,A$.

Proof:

(1) $A \supseteq FA \Rightarrow fA \supseteq fFA = \bigcup_{g \in F} fg\,A$.

(2) $A \subseteq FA \Rightarrow fA \subseteq fFA = \bigcup_{g \in F} fg\,A$. ○

In particular, for $f \in F^{<\omega}$, $fK = \bigcup_{g \in F} fg\,K$, showing how each nth-level image fK of K is composed of $(n+1)$th-level images of K. And using (1) note that . . .

__2.7__ __Corollary__ For $A \in \mathcal{K}X$, if F reduces A then for all $f,g \in F^{<\omega}$,

$$g \geq f \Rightarrow gA \subseteq fA .$$ ○

Finally, here's a simple result used later;

__2.8__ __Note__ For any sequence $\langle A_n \rangle$ in $\mathcal{K}X$ with all $A_n \subseteq K$ such that A_n intersects every nth-level image of K, $\langle A_n \rangle \to K$.

Proof:
For all n, $h(A_n, K) \leq r^n$ diam K using that each nth-level image of K has diameter $\leq r^n$ diam K. ○

3. The Code Map

By a *sequence* in F (without the term 'finite' appended or implied) we mean a function $\omega \to F$, i.e. an element of F^{ω}, and as with finite sequences we usually denote these with bold type $\boldsymbol{f}$ and $\boldsymbol{g}$. For $\boldsymbol{f} \in F^{\omega}$ and $n \in \omega$, $\boldsymbol{f} \restriction n$ denotes the restriction of $\boldsymbol{f}$ to n, a finite sequence of length n.

Recalling the remark after 2.6 pointing out how each nth-level image of K is composed of $(n+1)$th-level images of K, it follows that for each $\boldsymbol{f} \in F^{\omega}$, $\langle \boldsymbol{f} \restriction n\, K \rangle = \langle K, f_0 K, f_0 f_1 K, \ldots \rangle$ is a decreasing sequence of nonempty compact sets, whose diameters converge to zero as $\operatorname{diam} \boldsymbol{f} \restriction n\, K \leq r^n \operatorname{diam} K$, hence the intersection $\bigcap_n \boldsymbol{f} \restriction n\, K$ is a singleton whose element we now define as $k\, \boldsymbol{f}$. In this way we've defined a map $k : F^{\omega} \to K$, and we find that, naturally giving F^{ω} the product topology,

3.1 Proposition [Hu] [Wil] The map $k : F^{\omega} \to K$ is a continuous surjection.

Proof:

For $x \in K$, for any $f \in F^{<\omega}$ with $x \in fK$ there's $g \in F$ with $x \in fg\, K$ since $fK = \bigcup_{g \in F} fg\, K$; so by induction there's a sequence $\boldsymbol{f} \in F^{\omega}$ with $\forall n\ x \in \boldsymbol{f} \restriction n\, K$, so $x = k\, \boldsymbol{f}$ proving surjectivity of k. To prove continuity, suppose $\boldsymbol{f} \in F^{\omega}$ and $\boldsymbol{g} \in (F^{\omega})^*$ with $\boldsymbol{g} \in \mu \boldsymbol{f}$. Then $\forall n \in \omega$, $\boldsymbol{g} \restriction n \in \mu\, \boldsymbol{f} \restriction n$ so (using the topological aspect of 2.2) $\boldsymbol{g} \restriction n\, K^* \in \mu\, \boldsymbol{f} \restriction n\, K$, i.e. $\boldsymbol{g} \restriction n\, K^* \approx \boldsymbol{f} \restriction n\, K^*$, so as $k\, \boldsymbol{g} \in \boldsymbol{g} \restriction n\, K^*$ and $k\, \boldsymbol{f} \in \boldsymbol{f} \restriction n\, K^*$, ${}^{\circ}d(k\, \boldsymbol{g}, k\, \boldsymbol{f}) \leq \operatorname{diam} \boldsymbol{f} \restriction n\, K^* = \operatorname{diam} \boldsymbol{f} \restriction n\, K \leq r^n \operatorname{diam} K$. As $\langle r^n \operatorname{diam} K \rangle \to 0$, it follows that $k\, \boldsymbol{g} \approx k\, \boldsymbol{f}$ as required. ○

Since F^{ω} is a compact Hausdorff space, it follows moreover that the surjection k is *perfect*, i.e. closed and continuous with compact fibres ; and in particular it's a quotient map. Borrowing some terminology from [Ba], F^{ω} may be called code space since each $\boldsymbol{f} \in F^{\omega}$ can be thought of as a 'code' describing an element of K and all elements of K *have* such a code. Correspondingly, k may be called the code map.

Extending the operation of concatenation to the case of forming the infinite sequence $f\boldsymbol{g}$ from finite f and infinite $\boldsymbol{g}$ in the obvious way, we can state that . . .

3.2 Proposition [Hu] For $f \in F^{<\omega}$ and $\boldsymbol{g} \in F^{\omega}$, $f(k\, \boldsymbol{g}) = k\, (f \boldsymbol{g})$.

Proof:

$\forall n$, $f(k\, \boldsymbol{g}) \in f(\boldsymbol{g} \restriction n\, K) = (f(\boldsymbol{g} \restriction n))\, K = (f\boldsymbol{g}) \restriction (|f| + n)\, K \subseteq (f\boldsymbol{g}) \restriction n\, K$. ○

Note then that for $f \in F^{<\omega}$, $fK = k\,(f\, F^{\omega})$ where $f\, F^{\omega} = \{f\boldsymbol{g} \mid \boldsymbol{g} \in F^{\omega}\} = \{\boldsymbol{h} \in F^{\omega} \mid \boldsymbol{h} \text{ extends } f\}$.

We now consider the question of when each element of K has a *unique* code, i.e. when the code map k is injective (hence *bijective*). In fact this is equivalent to asking when k is a homeomorphism ;

3.3 Note k is a homeomorphism iff it's injective.

Proof:

F^{ω} is a compact Hausdorff space, so k is a homeomorphism iff it's a continuous bijection. We already know it's a continuous *sur*jection, so the result follows. ○

In the next proposition, when we refer to the family $\langle fK \mid f \in F^{n} \rangle$ of nth-level images of K being 'disjoint' we naturally mean that $f \neq g \Rightarrow fK$ and gK are disjoint.

3.4 Proposition The following are equivalent ;

(1) k is injective.

(2) $\langle fK \mid f \in F \rangle$ is disjoint and each $f \in F$ is injective on K.

(3) $\forall n \in \omega$, $\langle fK \mid f \in F^{n} \rangle$ is disjoint.

Proof:

(1) $\Rightarrow$ (2) : $\forall f,g \in F$, if $f \neq g$ then $f\, F^{\omega}$ and $g\, F^{\omega}$ are disjoint hence so are their images fK and gK under injective k. Secondly, each $f \in F$ is injective on K since for $g,h \in F^{\omega}$, $f(kg) = f(kh) \Rightarrow k(fg) = k(fh) \Rightarrow fg = fh \Rightarrow g = h \Rightarrow kg = kh$.

(2) $\Rightarrow$ (3) : This is quite easily seen and can be rigorously proved by induction on n.

(3) $\Rightarrow$ (1) : For distinct $f,g \in F^{\omega}$ there is n with $f{\restriction}n \neq g{\restriction}n$ so $f{\restriction}n\, K$ is disjoint from $g{\restriction}n\, K$, hence $kf \neq kg$ as required. ○

A further equivalence, easily seen from (3) , is that $\forall f \in F^{<\omega}$, fK is the *disjoint* union of the sets fgK $(g \in F)$.

The above proposition on injectivity of k is phrased in terms of K, making it more a 'structural' result than a practical test since we may not know what K is. However, we can use it to obtain the following more practical sufficiency condition ;

3.5 Corollary If there is $A \in \mathcal{K}X$ such that F reduces A and $\langle fA \mid f \in F \rangle$ is disjoint with all $f \in F$ injective on A, then k is injective.

Proof:

As F reduces A, $K \subseteq A$, hence $\langle fK \mid f \in F \rangle$ is disjoint and all $f \in F$ are injective on K, so k is injective. ○

For example, where $X = \mathbb{R}$ and F consists of the two maps contracting about 0 and 1 by factor 1/3, the set $A = [0,1]$ satisfies the conditions above, so k is injective. And in this case the attractor K, namely the intersection of the successive iterates $F^{n}A$ of A, is the classical Cantor Set, and the above shows that $K \cong 2^{\omega}$ since $F \cong 2 = \{0,1\}$.

Now here is a 'pointwise' result, on when $k\,f = k\,g$;

<u>3.6</u> <u>Proposition</u> For $f,g \in F^{\omega}$, $k\,f = k\,g \Leftrightarrow \forall n \;\; f{\upharpoonright}n\,K$ intersects $g{\upharpoonright}n\,K$.

Proof:

The forward implication is trivial, whilst for the reverse just apply 4.6 of Chapter 2 to the decreasing sequences $\langle f{\upharpoonright}n\,K\rangle$ and $\langle g{\upharpoonright}n\,K\rangle$ in $\mathcal{K}X$, which respectively have limits $\{k\,f\}$ and $\{k\,g\}$. ○

We conclude this section with a look at the limit action of the finite restrictions $f{\upharpoonright}n$ of a sequence $f \in F^{\omega}$. Bear in mind that as described in Appendix 2, the set $F^{\leq\omega} = F^{<\omega} \cup F^{\omega}$ of finite or infinite sequences in F naturally forms a topological space in which $F^{<\omega}$ and F^{ω} form subspaces with $F^{<\omega}$ open, and each $f \in F^{\omega}$ being the limit of its finite restrictions $f{\upharpoonright}n$.

<u>3.7</u> <u>Proposition</u> [Hu] For $f \in F^{\omega}$ and $A \in \mathcal{K}X$, $\langle f{\upharpoonright}n\,A\rangle \to \{k\,f\}$.

Proof:

The result is true for $A = K$ since as $\langle f{\upharpoonright}n\,K\rangle$ is decreasing we know (by 4.4 of Chapter 2) that it converges with respect to h to the intersection of its terms, which here is $\{k\,f\}$. The result for *arbitrary* $A \in \mathcal{K}X$ follows using that for all n, $h(f{\upharpoonright}n\,A,\, f{\upharpoonright}n\,K) \leq r^{n}\, h(A,K)$, with $\langle r^{n}\, h(A,K)\,\rangle \to 0$. ○

And since X is embedded in $\mathcal{K}X$, we have the following 'explicit' version of 1.8 ;

<u>3.8</u> <u>Corollary</u> For $f \in F^{\omega}$ and $x \in X$, $\langle f{\upharpoonright}n\,x\rangle \to k\,f$. ○

Rephrased, this says that $\langle \circ(f{\upharpoonright}n)\rangle$ converges pointwise (i.e. with respect to the topology we have on Lip_1X) to the constant function with value $k\,f$. Let us then define the <u>composition</u> $\circ f$ of f to be this constant function, and extending previous conventions, write $(\circ f)x$ as $f\,x$ and $(\circ f)A$ as $f\,A$. This gives an extension of the map $\circ : F^{<\omega} \to Lip_1X$ described in Section 2 to a map $\circ : F^{\leq\omega} \to Lip_1X$. Now as described in Appendix 2, $F^{\leq\omega}$ forms a 'partial' topological monoid in the sense that, where defined, the binary composition operation in $F^{\leq\omega}$ is associative and continuous. And with the obvious definition of a 'partial topological monoid action', we can now extend 2.1 to the following ;

<u>3.9</u> <u>Proposition</u> The composition map $\circ : F^{\leq\omega} \to Lip_1X$ is the unique partial topological monoid action of $F^{\leq\omega}$ on X such that the action of each $\langle f\rangle$ (for $f \in F$) is f.

Proof:

When $f\,g$ is defined, namely when f is finite, we have $\circ(f\,g) = (\circ f)\circ(\circ g)$, noting for the case where g is infinite that $\forall x$, $(f\,g)x = k\,(f\,g) = f(k\,g) = f(g\,x)$. This takes care of the algebraic aspect. Next we need to show continuity of the evaluation map

$F^{\leq\omega} \times X \to X$. In view of 2.1 and openness of $F^{<\omega}$, it remains only to check continuity at (f,x) for infinite f. So, take any $g \in \mu f$ and $\alpha \in \mu x$; since $fx = kf$, we must show $g\alpha \in \mu kf$. If g is $*$-infinite then $g\alpha = kg \in \mu kf$ by continuity of k. If g is $*$-finite then taking any $h \in (F^{\omega})^*$, as $d(\alpha, kh)$ is finite and g has infinitesimal ratio (namely $\leq r^{|g|}$, $|g|$ being infinite), $g\alpha \approx g(kh) = k(gh) \in \mu kf$ (noting $gh \in \mu f$) so $g\alpha \in \mu kf$ as required. Lastly, *uniqueness* (subject to the action of each $\langle f\rangle$ being f) follows from the uniqueness part of 2.1 and the necessity of the action of each *infinite* f being (by continuity) the limit of the actions of its finite restrictions. o

In particular then the composition map $\circ : F^{\leq\omega} \to Lip_1 X$ is continuous, which we can use in determining the closure of *Monoid* F in $C(X,X)$. Recalling that the constant map $X \to X$ with value $x \in X$ is denoted by c_x, then . . .

3.10 Proposition The closure of *Monoid* F in $C(X,X)$ is $(Monoid\ F) \cup \{c_x \mid x \in K\}$, which is compact.

Proof:

$F^{<\omega}$ is dense in compact $F^{\leq\omega}$, so taking their images under the continuous map $\circ : F^{\leq\omega} \to C(X,X)$, *Monoid* F is dense in compact $(Monoid\ F) \cup \{c_x \mid x \in K\}$ (noting that the image of F^{ω} is $\{c_{kf} \mid f \in F^{\omega}\} = \{c_x \mid x \in K\}$), which also shows that the latter is the closure of *Monoid* F in $C(X,X)$. o

For $A \in \mathcal{K}X$ it's natural to define an **ωth-level image** of A to be a set of the form fA where $f \in F^{\omega}$, namely $fA = \{kf\} = \lim \langle f\restriction n\, A\rangle$, the limit of the determined finite-level images of A. In this terminology, trivially K is the union of its ωth-level images, thus extending result 2.4. Moreover K is the union of the ωth-level images of *any* $A \in \mathcal{K}X$. And note that 2.5 now says that as n converges to ω, the set of nth-level images of A converges in $\mathcal{K}\mathcal{K}X$ to the set of ωth-level images of A.

4. Periodic Points

Note that $F^{<\omega} - \{\emptyset\}$ is the set of finite sequences in F of nonzero length. Using terminology from [Ba], we say $x \in X$ is periodic with respect to F if for some $f \in F^{<\omega} - \{\emptyset\}$, $fx = x$, i.e. x is the unique fixed point of $\circ f$, which we'll denote by 'fix f'. In other words the set of periodic points is $\{\text{fix } h \mid h \in \textit{Semigroup } F\}$, where ***Semigroup F*** is the closure of F under composition (the semigroup generated by F). And in fact, where f^{ω} denotes the sequence in which f is endlessly repeated, . . .

4.1 Proposition [Hu] For $f \in F^{<\omega} - \{\emptyset\}$, fix $f = k f^{\omega}$.
Proof:
Using 3.2, $f(k f^{\omega}) = k(f(f^{\omega})) = k f^{\omega}$. ∘

A sequence in F is naturally said to be *periodic* if it's of the form f^{ω} for some $f \in F^{<\omega} - \{\emptyset\}$. Note that for any $g \in F^{\omega}$, $g = \lim \langle (g \restriction n)^{\omega} \rangle$ in the space F^{ω}, so the periodic sequences are dense in F^{ω}.

4.2 Corollary The set of periodic points is the image under k of the set of periodic sequences in F.

In particular every periodic point is in K. And since the periodic sequences are dense in F^{ω}, their image under the continuous map k is dense in $k F^{\omega} = K$, i.e. . . .

4.3 Proposition [Hu] The periodic points are dense in K. ∘

Explicitly, . . .

4.4 Note [Hu] For $f \in F^{\omega}$, $kf = \lim \langle \text{fix } f \restriction n \rangle$.
Proof:
$f = \lim \langle (f \restriction n)^{\omega} \rangle$ so $kf = \lim \langle k(f \restriction n)^{\omega} \rangle = \lim \langle \text{fix } f \restriction n \rangle$. ∘

Finally, we point out the following and observe a minor corollary;

4.5 Note The map fix : *Semigroup F* $\to K$ is continuous.
Proof:
Just use A3.4 together with the fact that $r < 1$ is an upper bound on the Lipschitz ratios of the elements of *Semigroup F*. ∘

4.6 Corollary $K = \lim \langle \{\text{fix } f \mid f \in F^{n}\} \rangle_{n \geq 1}$.

Proof:
Each set $\{\text{fix } f \mid f \in F^{n}\}$ is in $\mathcal{K}X$ being the image of nonempty compact F^{n} under the continuous map $f \mapsto \text{fix}(\circ f)$ (a composition of *two* continuous maps), and is a subset of K intersecting every nth-level image of K (as fix $f \in fK$). As noted in 2.8, any such sequence converges to K. ∘

5. Connectedness of K

By earlier work, we have the following few results ;

5.1 Note If F is connected, so is K.

Proof:

If F is connected, so is F^{ω} and hence so is its image K under the continuous map k. o

5.2 Note If k is injective then

(1) K is connected $\Leftrightarrow$ F is connected.

(2) K is totally disconnected $\Leftrightarrow$ F is totally disconnected.

Proof:

Injectivity of k says $k : F^{\omega} \cong K$; and by general topological results, F^{ω} is connected iff F is, and F^{ω} is totally disconnected iff F is. o

5.3 Corollary For finite F with $|F| \geq 2$, if k is injective (in particular if F reduces some $A \in \mathscr{K}X$ such that $\langle fA \mid f \in F \rangle$ is disjoint with all $f \in F$ are injective on A) then K is totally disconnected. o

The above corollary is essentially also proved in Theorem 4.4 of [Ha]. Note incidentally that even when every element of F is a direct similitude of $X = \mathbb{R}^n$, K can be disconnected without being *totally* disconnected. As an example in $\mathbb{R}$ we could have $K = [0,1] \cup [2,3]$, formed as the union of four nonoverlapping copies of itself, each a quarter the size of K. Figure 12 of Section 1 shows another example, this time in $\mathbb{R}^2$.

The main result on when K is *connected* is due to Hata in the above-mentioned paper [Ha], and we now give a proof of this result along with some corollaries, first of all preparing with a lemma which generalizes a step used in Hata's proof. For sets A and B we may write $A \frown B$ for 'A intersects B'. We'll say a set $\mathscr{A}$ of sets is **interlinked** if for all $A,B \in \mathscr{A}$ there is a **linking chain** from A to B in $\mathscr{A}$, namely a finite sequence $C_1 \ldots C_n$ of elements of $\mathscr{A}$ such that $A = C_1 \frown \cdots \frown C_n = B$.

5.4 Lemma For $A \in \mathscr{K}X$, if F expands A and $\{fA \mid f \in F\}$ is interlinked then $\{fA \mid f \in F^n\}$ is interlinked for all n.

Proof:

First note that

(1) For all $f \in F^n$, $\{fgA \mid g \in F\}$ is interlinked.
For, whenever $g_1A \frown \cdots \frown g_mA$ then $fg_1A \frown \cdots \frown fg_mA$.

(2) For all $f,g \in F^n$ with $fA \frown gA$, $\{fhA \mid h \in F\}$ 'links with' $\{ghA \mid h \in F\}$ in the sense that some fhA intersects some ghA. This is just because $fA \subseteq \bigcup_{h \in F} fhA$ and $gA \subseteq \bigcup_{h \in F} ghA$ (by 2.6(2)).

We are now ready to prove the lemma by induction on n. The hypothesis is true for $n = 1$ by assumption (and also trivially true for $n = 0$). Now let $n \geq 1$ and assume true for n. Take any $(n+1)$th-level images $f_1 g_1 A$ and $f_2 g_2 A$ of A ($f_1, f_2 \in F^n$ and $g_1, g_2 \in F$). By the hypothesis for n there is a linking chain $h_1 A \wedge \cdots \wedge h_m A$ from $f_1 A$ to $f_2 A$ in $\{h A \mid h \in F^n\}$ (with $h_1 = f_1$ and $h_m = f_2$); now by (1) each set $\{h_i g A \mid g \in F\}$ is interlinked, and by (2) each $\{h_i g A \mid g \in F\}$ is 'linked with' $\{h_{i+1} g A \mid g \in F\}$, hence as required there is a linking chain of $(n+1)$th-level images of A from $h_1 g_1 A = f_1 g_1 A$ to $h_m g_2 A = f_2 g_2 A$, proving the hypothesis for $n+1$. $\circ$

Now to Hata's result, with the slight addition of (4) by the author (as well as the case for *infinite* F). The proof is closely that of Hata's, who uses the special case of the above lemma in which $A = K$;

5.5 Proposition [Ha] For finite F the following are equivalent, whilst for infinite F we have the implications (3) $\Leftrightarrow$ (4) $\Rightarrow$ (1).

(1) K is connected.

(2) K is connected and locally connected.

(3) $\{fK \mid f \in F\}$ is interlinked.

(4) $\{fK \mid f \in F^n\}$ is interlinked for all n.

Proof:

Since F expands K the last lemma gives (3) $\Leftrightarrow$ (4). Now if (4) holds then bearing in mind that for each n K is the union of its nth-level images and that these have diameters $\leq r^n \operatorname{diam} K$, it follows that K is 'chain-connected' (i.e. $\forall x,y \in K$ $\forall \epsilon > 0$ there's an 'ϵ-chain from x to y in K', i.e. a finite sequence in K starting at x and ending at y such that successive points are distance at most ϵ apart), and since K is compact this is equivalent to K being connected. This establishes (4) $\Rightarrow$ (1). For the remaining implications assume F is finite;

(1) $\Rightarrow$ (3) : Let $f \in F$ and $G = \{g \in F \mid$ there's a linking chain from fK to gK in $\{hK \mid h \in F\}\}$. If $G \neq F$ then for all $g \in G$ and $h \notin G$ gK is disjoint from hK, thus $\bigcup_{g \in G} gK$ is disjoint from $\bigcup_{h \notin G} hK$, and since these are nonempty closed sets K is not connected, contradiction; so $G = F$ as required.

(1) $\Rightarrow$ (2) : As K is connected so is each nth-level image of K (being a continuous image of K), and since K is their union we have that $\forall \epsilon > 0$ K is the union of finitely many connected sets of diameter $\leq \epsilon$; hence by a general result K is locally connected. $\circ$

As also noted by Hata, (2) in other words says that K is a locally connected continuum (a 'continuum' being a nonempty connected compact space), which by the Hahn-Mazurkiewicz theorem is equivalent to K being a continuous image of $[0,1]$.

Also note that the implication '(1) ⇏ (2)' cannot hold for *arbitrary* F, simply by 1.2 and the fact that there can exist connected $A \in \mathcal{K}X$ without A being *locally* connected. Hata's result easily yields the following sufficient condition for K to be connected ;

__5.6__ __Corollary__ If some $A \in \mathcal{K}X$ is expanded by F with $\{fA \mid f \in F\}$ interlinked, then K is connected.

Proof:

We know $A \subseteq K$, so $\{fK \mid f \in F\}$ is interlinked too. o

In particular, as $\{\mathrm{fix}\, f \mid f \in F\}$ is expanded by F, and is compact being the image of F under the continuous map fix : *Semigroup* $F \to K$ in 4.5, then . . .

__5.7__ __Corollary__ Letting $A = \{\mathrm{fix}\, f \mid f \in F\}$, if $\{fA \mid f \in F\}$ is interlinked then K is connected. o

For example the above applies to show that the Sierpinski Gasket is connected. Alternatively 5.6 could have been used, by taking A to be the boundary of the triangle involved. Here's a further sufficient condition for K to be connected. Since the set of connected elements of $\mathcal{K}X$ is (by 4.3 of Chapter 1) closed in $\mathcal{K}X$, and $K = \lim \langle F^n A \rangle$ for all $A \in \mathcal{K}X$, then . . .

__5.8__ __Note__ If there's $A \in \mathcal{K}X$ such that all iterates $F^n A$ of A under $\bigcup F$ are connected, then K is connected. o

__5.9__ __Corollary__ If there's connected $A \in \mathcal{K}X$ such that $\{fA \mid f \in F^n\}$ is interlinked for all n, then K is connected.

Proof:

As $\{fA \mid f \in F^n\}$ is interlinked then since each fA is connected (being a continuous image of A) their union $F^n A$ is connected (by a general result). o

Note that in the case where F *expands* connected $A \in \mathcal{K}X$ we can replace the condition in the above by the condition that $\{fA \mid f \in F\}$ be interlinked, thus giving an alternative path to 5.6. On the other hand, if F is finite and *reduces* connected $A \in \mathcal{K}X$, then K is connected *iff* $\{fA \mid f \in F^n\}$ is interlinked for all n, since if K is connected then each $\{fK \mid f \in F^n\}$ is interlinked hence so is each $\{fA \mid f \in F^n\}$ as in general $fK \subseteq fA$.

6. Regularity and Residuality of K

In this section we show that subject to the 'homeomorphism condition' below, K is either regular or residual (see 'Preliminaries' for a reminder on these terms).

Throughout let hF denote the set of elements of F which are *homeomorphisms* of X. The **homeomorphism condition** on F, expressing that the elements of F which are *not* homeomorphisms are nevertheless tied down to some extent by those which *are*, is that $F - hF \subseteq \overline{Semigroup\ hF}$ (implicitly demanding that hF be nonempty), equivalently that $Semigroup\ hF$ is dense in $Semigroup\ F$ (noting that if the first condition holds, then $F \subseteq \overline{Semigroup\ hF}$, so as $\overline{Semigroup\ hF}$ is a semigroup, $Semigroup\ F \subseteq \overline{Semigroup\ hF}$, i.e. $Semigroup\ hF$ is dense in $Semigroup\ F$). In the case where *every* element of F is a homeomorphism, the condition is of course trivially satisfied and we can expect the general theory concerned with K to be simpler than it otherwise might. The main reason we widen the scope of the work below from this simpler situation to the situation in which merely the homeomorphism condition is required, is for the sake of an application at the end of Section 4 of Chapter 4 when we show that certain invariant sets may be 'view self-similar'.

6.1 Note If F satisfies the homeomorphism condition,

(1) $\{\text{fix } f \mid f \in Semigroup\ hF\}$ is dense in K.

(2) For any $x \in K$, any neighbourhood N of x, and any bounded $B \subseteq X$, there's $f \in Semigroup\ hF$ with $fB \subseteq N$.

Proof:

(1) $Semigroup\ hF$ is dense in $Semigroup\ F$, and the map $\text{fix} : Semigroup\ F \to K$ is continuous by 4.5, so $\{\text{fix } f \mid f \in Semigroup\ hF\}$ is dense in $\{\text{fix } f \mid f \in Semigroup\ F\}$; and as the latter is dense in K (4.3) so is the former.

(2) By (1) it suffices to prove this for the case where $x = \text{fix } g$ for some $g \in Semigroup\ hF$. Being bounded, B is a subset of some ball on x, and if we apply the contraction g repeatedly then the ball will be mapped inside N; thus we can take f to be some power of g. ○

The homeomorphism condition is of little interest in the case of *finite* F since hF has the same attractor as F, thus allowing us to reduce things to the case where all the contractions involved are homeomorphisms ;

6.2 Corollary If F is finite and satisfies the homeomorphism condition, $K_F = K_{hF}$.

Proof:

Being nonempty and finite, hF is admissible. And by the last note, K_F is the closure of $\{\text{fix } f \mid f \in Semigroup\ hF\}$; but so is K_{hF} of course. ○

6.3 Lemma If F satisfies the homeomorphism condition, the following are equivalent for any $A \in \mathcal{K}X$ reduced by F;

(1) $K \cap \underline{A}$ is dense in K.

(2) K intersects $\underline{A}$.

(3) For some $f \in$ *Semigroup hF*, fix $f \in \underline{A}$.

(4) For some $f \in$ *Semigroup hF*, $fA \subseteq \underline{A}$.

Proof:

(1) ⇒ (2) : Trivial.

(2) ⇒ (1) : Let $x \in K$, and N be a neighbourhood of x. As A is bounded there's $f \in$ *Semigroup* hF with $fA \subseteq N$, so (as $K \cap \underline{A} \neq \emptyset$) $\emptyset \neq f(K \cap \underline{A}) \subseteq fA \subseteq N$; and $f(K \cap \underline{A}) = fK \cap f\underline{A} \subseteq K \cap \underline{A}$, so $K \cap \underline{A}$ intersects N.

(2) ⇔ (3) : Since {fix f | $f \in$ *Semigroup hF* } is dense in K, it intersects $\underline{A}$ iff K does.

(2) ⇒ (4) : Let $x \in K \cap \underline{A}$. Then as $\underline{A}$ is a neighbourhood of x and A is bounded, there's $f \in$ *Semigroup hF* with $fA \subseteq \underline{A}$.

(4) ⇒ (2) : Then since $fK \subseteq fA \subseteq \underline{A}$, K intersects $\underline{A}$. ○

We now have the main result ;

6.4 Proposition If F satisfies the homeomorphism condition, K is regular or residual.

Proof:

If K is not residual, i.e. $\underline{K} \neq \emptyset$, then K intersects $\underline{K}$ so by 6.3 $K \cap \underline{K}$ is dense in K, i.e. $\underline{K}$ is dense in K, i.e. K is regular. ○

The following result gives a sufficient condition for K to be residual in the case of $X = \mathbb{R}^n$. For $A \subseteq \mathbb{R}^n$, ***dim A*** denotes the topological dimension of A (for the definition of which, see for example [HW]).

6.5 Proposition Let $X = \mathbb{R}^n$ and F be finite with $F = hF$.

Then if for all distinct $f,g \in F$ $dim\,(fK \cap gK) \leq n\text{-}2$, equivalently if some $A \in \mathcal{K}X$ is reduced by F such that for all distinct $f,g \in F$ $dim\,(fA \cap gA) \leq n\text{-}2$, K is residual.

Proof:

The equivalence of the two conditions is easily seen, using that if the second holds then so will the first as $K \subseteq A$. Assume now the condition on K. By induction on n (using $F = hF$) it follows that for all $n \geq 1$, for all distinct $f,g \in F^n$ $dim\,(fK \cap gK) \leq n\text{-}2$.

Suppose now for a contradiction that $\underline{K} \neq \emptyset$; let B be a closed ball with $B \subseteq K$. Let $0 < \epsilon < \text{diam } B$, and take n such that $r^n \text{ diam } K \leq \epsilon$ so that all nth-level images of K have diameter $\leq \epsilon$.

Letting $\mathcal{A} = \{B \cap fK \mid f \in F^n \text{ and } B \cap fK \neq \emptyset\}$, $\mathcal{A}$ is a finite set of closed-in-B proper subsets of B such that for all distinct $A_1, A_2 \in \mathcal{A}$ $dim\,(A_1 \cap A_2) \leq n\text{-}2$, and such that $B = \bigcup \mathcal{A}$. But B *cannot* be the union of such a set $\mathcal{A}$, the following proof of which stems from a helpful suggestion of Dona Strauss.

For, letting I be the set of intersection points among the elements of $\mathscr{A}$, namely $I = \bigcup\{A_1 \cap A_2 \mid A_1 \text{ and } A_2 \text{ are distinct elements of } \mathscr{A}\}$, we have $dim\, I \leq n\text{-}2$ (by the 'sum theorem' in [HW], page 30), and I disconnects B in the sense that $B - I$ is not connected. For where $\mathscr{A}' = \{A - I \mid A \in \mathscr{A} \text{ and } A - I \neq \emptyset\}$, $\mathscr{A}'$ is a finite partition of $B - I$ into mutually disjoint closed-in-$(B - I)$ sets, with each $A - I$ being a *proper* subset of $B - I$ (otherwise we'd have $B - I = A - I \subsetneq A$ giving (as $I = \emptyset$, as $dim\, I < n$; see [HW], page 44) $B = \overline{B - I} \subsetneq A$ contradicting $A \subset B$). This is a contradiction since B cannot be disconnected by a subset of dimension $\leq n\text{-}2$ (see [HW], page 48). ○

6.6 Corollary Let $n \geq 2$ and $X = \mathbb{R}^n$, and F be finite with $F = hF$.
Then if for all distinct $f,g \in F$ $fK \cap gK$ is finite, equivalently if some $A \in \mathscr{K}X$ is reduced by F such that for all distinct $f,g \in F$ $fA \cap gA$ is finite, K is residual. ○

In particular the above corollary shows that the 'nested fractals' on which Lindstrøm has constructed Brownian motion (see [Li,2]) are all residual, in particular the Sierpinski Gasket (take A to be the triangle involved for example). The condition $F = hF$ in 6.5 cannot be omitted, because the induction used in the first part of the proof will break down. Here's an example of this in which $X = \mathbb{R}^2$. Let m_1 be the operation of 'folding' $\mathbb{R}^2$ about the x-axis so that the upper half-plane maps to the lower half-plane (explicitly, m_1 is constant on the the lower half-plane, whilst reflecting the *remaining* points in the x-axis). Likewise, let m_2 be the operation of folding $\mathbb{R}^2$ about the y-axis so that the right half-plane maps to the left. Let m_3 contract about (-2,-2) by factor 1/2, and let $f_1 = m_3 \circ m_2 \circ m_1$, which has Lipschitz ratio 1/2. Let $A = [-2, 2]^2$ and note that $f_1A = [-2,-1]^2$. Follow f_1 by the obvious translations to get f_2 up to f_8 such that where $F = \{f_1, \ldots, f_8\}$, FA is the shaded set K in the diagram. Note that K is invariant under F, so K is the attractor of F,

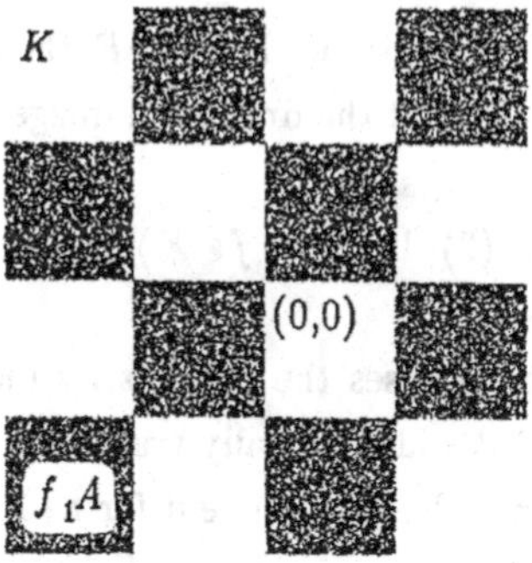

and K is *not* residual, despite the fact that $fK \cap gK$ is finite (hence of dimension 0) for all distinct $f,g \in F$. The trouble, stemming from the fact that the elements of F are not homeomorphisms, lies in the fact that the higher level images of K need *not* have the same intersection properties as the first-level images; for example, $f_1 f_1 K = [-2,-1]^2 = f_1 f_8 K$, which is of course *not* of dimension 0.

A more widely applicable sufficient condition for residuality of K will follow from the work of the next section, in 7.13.

7. Tiling of K

In this section we look at the internal composition of K in terms of the way it's formed as the union of its first-level images, isolating a particularly tidy situation. Introducing some terminology first, we say a family $\langle A_i \mid i \in I \rangle$ of subsets of a space Y is a **tiling** of Y if it's a nonoverlapping cover of Y by nonempty regular closed sets. The term 'nonoverlapping' is being used in a formal sense here and is described in Appendix 5 with which familiarity will be henceforth assumed. We'll say K is **tiled** with respect to F if $\langle fK \mid f \in F \rangle$ is a tiling of K. For example, the Sierpinski Gasket is tiled by the usual F producing it. Of course, here no fK is regular in X, but that is not required by the definition; only regularity in K (which is weaker). The definition of K being tiled with respect to F is concerned only with the *intrinsic* nature of K. It should also be pointed out that . . .

7.1 Note If K is tiled with respect to F, F must be finite.

Proof:

$\{fK \mid f \in F\}$ is compact in the Vietoris topology on $\mathscr{K}K$ (being a continuous image of F), hence by A5.4 it must be finite, and it follows in turn that F must be finite since the map $f \mapsto fK$ is injective (for if $f \neq g$ then fK and gK are nonoverlapping nonempty closed regular sets in K hence $fK \neq gK$). ○

The next result in particular applies to K, showing that as long as every element of F is a homeomorphism, both regularity and essential disjointness in K of its first-level images carry through to its nth-level images in general ;

7.2 Proposition If $F = hF$ then for any $A \in \mathscr{K}X$ reduced by F,

(1) If the first-level images of A are regular in A, so are the nth-level images for all n.

(2) If $\langle fA \mid f \in F \rangle$ is nonoverlapping in A, so is $\langle fA \mid f \in F^n \rangle$ for all n.

Proof:

In both cases the proof is by induction on n. We're given the hypothesis for $n = 1$ (and it's also trivially true for $n = 0$), so now assume $n \geq 1$ and the hypothesis is true for n. We now prove it for $n+1$;

(1) Take any element of F^{n+1}, say gf where $g \in F^n$ and $f \in F$. As $(\circ g)\restriction A : A \cong gA$ and fA is regular in A, then gfA is regular in gA hence also in A (as gA is regular in A). So the hypothesis is proved for $n+1$.

(2) Take any distinct elements of F^{n+1}, say gf_1 and hf_2 where $g, h \in F^n$ and $f_1, f_2 \in F$. If $g = h$ then as $(\circ g)\restriction A : A \cong gA$ with f_1A and f_2A nonoverlapping in A, gf_1A and $gf_2A = hf_2A$ are nonoverlapping in gA hence also in A as required. If $g \neq h$ then as $gf_1A \subseteq gA$ and $hf_2A \subseteq hA$ with gA and hA nonoverlapping in A, gf_1A and hf_2A are nonoverlapping in A too. So the hypothesis holds for $n+1$. ○

7.3 Corollary If $F = hF$ and K is tiled with respect to F, then $\langle fK \mid f \in F^n \rangle$ is a tiling of K for all n. ○

A fairly trivial case of K being tiled with respect to F is the following ;

7.4 Proposition If F is finite and $\langle fK \mid f \in F \rangle$ is disjoint, K is tiled with respect to F.

Proof:
As $\langle fK \mid f \in F \rangle$ is disjoint, trivially it's nonoverlapping in K, and each fK is clopen in K (using finiteness of F) hence regular in K. ○

In the above result, as long as $|F| \geq 2$ K will be disconnected of course (*totally* disconnected if $F = hF$; recall 5.3). Towards establishing a more *generally* applicable sufficient condition for K to be tiled with respect to F we introduce the following definition. The **compact set condition** on F is that there exists a regular nonempty compact set A (or 'body' for short; see 'Preliminaries') reduced by F with $\langle fA \mid f \in F \rangle$ nonoverlapping. K is a subset of any such set of course, hence $\langle fK \mid f \in F \rangle$ will be nonoverlapping in X too, but this need not be saying much ; for if every fK is residual, *trivially* $\langle fK \mid f \in F \rangle$ is nonoverlapping in X ; and it certainly does not follow that $\langle fK \mid f \in F \rangle$ is nonoverlapping in K. Related to the compact set condition is a condition appearing in the literature in various forms, namely the **open set condition** that there exists a body-interior D closed under all $f \in F$ with $\langle fD \mid f \in F \rangle$ disjoint. Indeed, bearing in mind that the bodies correspond *bijectively* with the body-interiors via the mutually inverse maps $A \mapsto \underline{A}$ and $D \mapsto \overline{D}$, we have the following ;

7.5 Note If $F = hF$ then

(1) If the compact set condition holds with A, the open set condition holds with $\underline{A}$.

(2) If the open set condition holds with D, the compact set condition holds with $\overline{D}$.

(3) The compact set condition and the open set condition are equivalent.

Proof:
(1) As A is closed under all $f \in F$ so is $\underline{A}$. And as $\langle fA \mid f \in F \rangle$ is nonoverlapping then $\langle \underline{fA} \mid f \in F \rangle$ is disjoint, i.e. $\langle f\underline{A} \mid f \in F \rangle$ is disjoint.

(2) As D is closed under all $f \in F$ so is $\overline{D}$. And as $\langle fD \mid f \in F \rangle$ is disjoint with each fD open, $\langle \overline{fD} \mid f \in F \rangle$ is nonoverlapping, i.e. $\langle f\overline{D} \mid f \in F \rangle$ is nonoverlapping.
(3) By (1) and (2). ○

In the case $F = hF$ the open set condition is also equivalent to merely the existence of a nonempty subcompact open set V closed under all $f \in F$ with $\langle fV \mid f \in F \rangle$ disjoint, since F will then satisfy the compact set condition with $\overline{V}$ (and hence the open set condition with the regularization $\underline{(\overline{V})}$ of V). It is usually this statement of the open set condition that has appeared in the literature (e.g. see [Fa] and [Li,1] or [Li,2]), though with 'subcompact' equivalently replaced by 'bounded' since the interest has only been in $\mathbb{R}^n$. In [Hu] however, the requirement of boundedness is omitted altogether. The literature seems to have dealt *solely* in the open set condition, but for our purposes the *compact* set condition seems more natural. A few notes on the condition follow.

<u>7.6</u> <u>Note</u> If $F = hF$ and F satisfies the compact set condition, F must be finite.

Proof:

Suppose F satisfies the compact set condition with A. Then $\{fA \mid f \in F\}$ is a cover of FA by nonempty compact sets with each fA regular in FA (being regular in X) and with $\{fA \mid f \in F\}$ nonoverlapping in FA (being nonoverlapping in X with each fA regular in X (see A5.1)) ; thus it's a tiling of FA. And with reasoning similar to that in 7.1, it follows that F must be finite. ∘

<u>7.7</u> <u>Note</u> If $F = hF$ and F satisfies the compact set condition with A, then for all n $\langle fA \mid f \in F^n \rangle$ is nonoverlapping with each fA regular.

Proof:

As A is regular so is each $fA = (\circ f)A$ since $\circ f$ is a homeomorphism. And as $\langle fA \mid f \in F \rangle$ is nonoverlapping in X, it's equivalently nonoverlapping in A as each fA is regular; so by 7.2(2) $\langle fA \mid f \in F^n \rangle$ is nonoverlapping in A, equivalently in X. ∘

<u>7.8</u> <u>Note</u> If $F = hF$ then $\bigcup F$ preserves regularity, i.e. if $A \in \mathcal{K}X$ is regular so is FA.

Proof:

If $A \in \mathcal{K}X$ is regular, so is each first-level image fA of A, hence so is the union FA by the simple result that any union of regular closed sets which is *closed* is regular. ∘

<u>7.9</u> <u>Note</u> If $F = hF$ and F satisfies the compact set condition with A, it also satisfies it with FA.

Proof:

By the last note FA is regular, and as A is reduced by F so is FA. Lastly, as $\langle fA \mid f \in F \rangle$ is nonoverlapping so is $\langle fFA \mid f \in F \rangle$ since each $fFA \subseteq fA$. ∘

<u>7.10</u> <u>Lemma</u> If $F = hF$ and F satisfies the compact set condition with A, then $\forall f \in Monoid\, F$, $K \cap \underline{fA} = fK \cap \underline{fA} = f(K \cap \underline{A})$.

Proof:

Let $f = \circ f$ where $f \in F^n$. Since $\langle gA \mid g \in F^n \rangle$ is nonoverlapping (by 7.7) then for $g \in F^n - \{f\}$, gA is disjoint from $\underline{fA}$, so gK is disjoint from $\underline{fA}$ too. So as K is the union of its nth-level images, $K \cap \underline{fA} = fK \cap \underline{fA}$. Lastly, as $\circ f$ is a homeomorphism, $fK \cap \underline{fA} = fK \cap f\underline{A} = f(K \cap \underline{A})$. ∘

We now have the first main result of this section;

7.11 Proposition If $F = hF$ and F satisfies the compact set condition with some A such that K intersects $\underline{A}$, then K is tiled with respect to F.

Proof:

Taking $f \in F$, let $V = K \cap \underline{fA}$, equivalently $V = fK \cap \underline{fA} = f(K \cap \underline{A})$. Then V is an open-in-K subset of fK, and is dense in fK as $K \cap \underline{A}$ is dense in K (by 6.3(1)). So fK is regular in K. And for $g \in F - \{f\}$, fK is disjoint from $\underline{gK}_K$ ($\underline{\ \ }_K$ denoting 'interior in K'), otherwise the set V (being dense in fK) would intersect $\underline{gK}_K$, implying that $\underline{fA}$ intersects gA (as $V \subseteq \underline{fA}$ and $\underline{gK}_K \subseteq gK \subseteq gA$), contradicting that fA and gA are nonoverlapping. o

Whether or not the condition that K intersects $\underline{A}$ can be *omitted* in the above is at present unknown to me. Certainly K need *not* intersect the interior of every A satisfying the compact set condition; i.e. K *could* be a subset of the boundary ∂A of A.

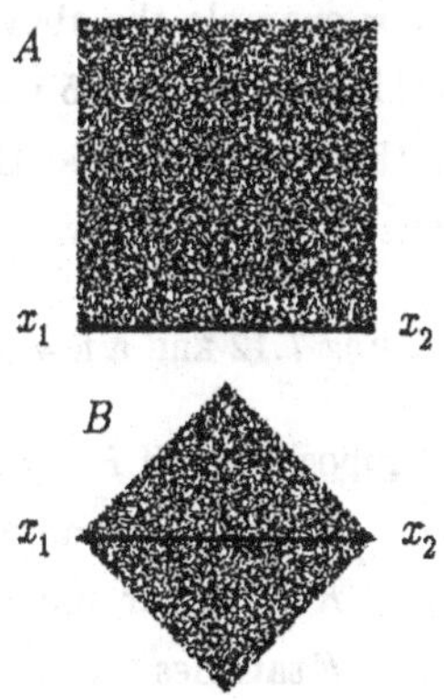

As an example of this let $X = \mathbb{R}^2$, $F = \{f_1, f_2\}$ where f_1 and f_2 contract by factor 1/2 about $x_1 \neq x_2$ respectively, so $K = [x_1, x_2]$, and let A be a rectangle with base K. Although $K \subseteq \partial A$, K is nevertheless tiled with respect to F. Indeed, there *is* a set B satisfying the compact set condition with K intersecting $\underline{B}$; for example the square with diagonal K. Also note that in this example there's a *one*-dimensional affine subspace of $\mathbb{R}^2$ closed under the elements of F (within which K therefore lies); this could be relevant. However, a more fundamental open question is the following. If F is finite and $F = hF$, and $\langle fK \mid f \in F \rangle$ is nonoverlapping in K, does it follow *automatically* that the sets fK are regular in K, i.e. that K is tiled with respect to F? For $|F| = 2$ at least, the answer is yes. See note 10.11 for some remarks on the matter.

Our second main result is concerned with *regular* K being tiled with respect to F;

7.12 Proposition If $F = hF$ and K is regular, then

(1) K is tiled with respect to $F \Leftrightarrow \langle fK \mid f \in F \rangle$ is nonoverlapping.

(2) F can satisfy the compact set condition only with K; and it does so iff K is tiled with respect to F.

Proof:

(1) The first-level images of K are regular in X, hence (as K is regular) also in K. So K is tiled with respect to F iff $\langle fK \mid f \in F \rangle$ is nonoverlapping in K. But since the sets fK are regular, $\langle fK \mid f \in F \rangle$ is nonoverlapping in K iff $\langle fK \mid f \in F \rangle$ is nonoverlapping in X.

(2) Suppose F satisfies the compact set condition with A. We must show $A = K$, i.e. $A \subseteq K$. As $\underline{K} \neq \emptyset$, there's $f \in$ *Semigroup* F with $f\underline{A} \subseteq K$ (K is a neighbourhood of some

$x \in K$, and we can apply 6.1(2) to $B = \underline{A}$), so $f\underline{A} = K \cap f\underline{A} = K \cap \underline{fA} = f(K \cap \underline{A})$ (using 7.10 for the last equality) giving $\underline{A} = K \cap \underline{A}$, i.e. $\underline{A} \subseteq K$, so $A \subseteq K$ as required. Lastly, by (1) and the regularity in K of the first-level images of K, F satisfies the compact set condition with K iff K is tiled with respect to F. ○

As a consequence we have the following quite widely applicable sufficient condition for residuality of K ;

7.13 Corollary If $F = hF$ and F satisfies the compact set condition with some A such that $FA \neq A$ (i.e. $A \neq K$), then K is residual.

Proof:

By 6.4 K is either regular or residual. But by 7.12 it cannot be regular as F satisfies the compact set condition with some $A \neq K$. ○

For example the above shows that the well-known Sierpinski Carpet is residual (something for which 6.5 was too weak), by taking A to be the square involved. It also shows that the Sierpinski Gasket is residual (though this also followed from the more specialized result 6.6) by taking A to be the triangle involved.

Using 7.12 and a few earlier results we observe also the following equivalences ;

7.14 Proposition If $F = hF$, the following are equivalent ;

(1) K is regular and is tiled with respect to F.

(2) K is regular and F satisfies the compact set condition.

(3) F satisfies the compact set condition with a unique A.

(4) F satisfies the compact set condition with a least A.

(5) F satisfies the compact set condition with K.

Proof:

(1) ⇒ (2) : By 7.12(2) F satisfies the compact set condition with K.

(2) ⇒ (3) : By 7.12(2) again.

(3) ⇒ (4) : Trivial.

(4) ⇒ (5) : If F satisfies the compact set condition with least A, then since it also satisfies it with FA (by 7.9) we have $A \subseteq FA$, hence $FA = A$ so $A = K$.

(5) ⇒ (1) : By 7.11, taking $A = K$. ○

Finally we point out the connection of the above subject matter with 'reptiles', which have been known in the literature of tilings for some time, the term being short for 'replicating tile'. A *reptile* (see [Mar] for example) is a polygon in $\mathbb{R}^2$ which is tiled by smaller mutually congruent copies of itself; in our terms it's a polygon of the form K_F for some finite set F of similitudes of $\mathbb{R}^2$ having the same scale factor, such that K_F is tiled with respect to F. By an existence theorem for tilings (a general form of which is proved in the latter half of Appendix 5 using nonstandard analysis and various hyperspace work), any reptile can be used to tile $\mathbb{R}^2$. The condition of being a *polygon* is superfluous in this respect, as is the dimension of the space; we can generalize for example as follows. By an *isometric copy* of $K = K_F$ we'll mean an image of K under an isometry of the space, whilst by a *direct* isometric copy we'll mean an image under a *direct* isometry.

7.15 Proposition For any finite nonempty set F of [direct] similitudes of $\mathbb{R}^n$ with common scale factor $r < 1$ such that K is regular and tiled with respect to F, $\mathbb{R}^n$ can be tiled by [direct] isometric copies of K.

Proof:
(Note that in the following, 'copy' will mean 'isometric copy' or 'direct isometric copy' according to which version of the result we're proving.)

By an existence theorem (e.g. A5.13 with $\mathscr{A} = \{K\}$) it suffices to show that every bounded subset of $\mathbb{R}^n$ can be covered by mutually nonoverlapping copies of K. Taking any $x \in \underline{K}$ let g be the dilation about x by factor $1/r$. Then for all n, letting I_n be the set of nth-level images of K, $g^n I_n$ is a set of mutually nonoverlapping copies of K with union $g^n K$. Since every bounded set is covered by a set of the form $g^n K$, the proof is complete. ○

In connection with this result, see also note 7.8 of Chapter 4, concerned with obtaining *explicit* examples of tilings of X by isometric copies of K.

8. Continuity of the Attractor Map

Recall that for each admissible set F of contractions of X we have the 'attractor' K_F of F. We thus have a map $K: Admis\ X \to \mathscr{K}X$ taking each F to its attractor K_F. We shall call this the **attractor map**. $Admis\ X$ is a subset of $\mathscr{K}C(X,X)$, and giving the latter the Vietoris topology induced from the compact-uniform topology on $C(X,X)$, $Admis\ X$ becomes a topological space, namely a subspace of $\mathscr{K}C(X,X)$, and we are now free to consider the possible continuity of the attractor map. The main result of this section is that . . .

8.1 Proposition If X is locally compact, the attractor map is continuous.

Proof:

The attractor map is the composition of $\bigcup : Admis\ X \to Contrac\,\mathscr{K}X$ followed by $\text{fix} : Contrac\,\mathscr{K}X \to \mathscr{K}X$. And these are both continuous, the first by 6.6 of Chapter 2 (local compactness not being needed here), and the second by A3.2 applied to $\mathscr{K}X$ (noting that $\mathscr{K}X$ is locally compact as X is). o

For non-locally-compact X the attractor map need *not* be continuous, due to the fact that $\text{fix} : Contrac\ X \to X$ need not be. For if $g \in \mu f$ with $\text{fix}\ g \notin \mu\ \text{fix}\ f$, then taking $F = \{f\}$ and $G = \{g\}$ we have $G \in \mu F$ in $Admis\ X$, but since $K_G = \{\text{fix}\ g\}$ and $K_F = \{\text{fix}\ f\}$ then $K_G \notin \mu\ K_F$. However, by placing an upper bound u on the Lipschitz ratios allowed, we can obtain the following restricted result. For $u < 1$ let $\boldsymbol{Admis_u} X = \{F \in Admis\ X \mid \forall f \in F\ r_f \le u,\ \text{i.e.}\ \bigvee_{f \in F} r_f \le u\}$, and for any metric space Y let $\boldsymbol{Contrac_u} Y = \{f \in Contrac\ Y \mid r_f \le u\}$. Then . . .

8.2 Proposition For $u < 1$, $K: Admis_u X \to \mathscr{K}X$ is continuous.

Proof:

Similar to that of 8.1 except taking the first map to be $\bigcup : Admis_u X \to Contrac_u \mathscr{K}X$ (noting that for $F \in Admis_u X$, $\bigcup F \in Contrac_u \mathscr{K}X$ as $r_{\bigcup F} \le \bigvee_{f \in F} r_f \le u$) and the second map to be $\text{fix} : Contrac_u \mathscr{K}X \to \mathscr{K}X$, which by A3.4 is continuous (*without* requiring local compactness of $\mathscr{K}X$). o

From the above two results we can obtain continuity results closer to the type appearing in the literature (namely in [Ha], [Li,1] and [Ba]), dealing with finite *families* $F = \langle f_1, \ldots, f_n \rangle \in (Contrac\ X)^n$ of contractions of X. Each such family determines the nonempty finite *set* $\{f_1, \ldots, f_n\}$ of contractions of X, whose attractor we shall naturally denote by K_F (calling this the attractor of F). We thus have another 'attractor map' $K: (Contrac\ X)^n \to \mathscr{K}X$, and since the map of 'set formation' sending each $\langle f_1, \ldots, f_n \rangle$ to $\{f_1, \ldots, f_n\}$ is easily seen to be continuous, composition of continuity gives, respectively using 8.1 and 8.2, that . . .

8.3 Corollaries

(1) If X is locally compact, $K: (Contrac\ X)^n \rightarrow \mathcal{K}X$ is continuous.

(2) For any $u < 1$, $K: (Contrac_u X)^n \rightarrow \mathcal{K}X$ is continuous. ○

An example of the continuity given in (1) is illustrated below, where we see a sequence of invariant sets belonging to the family described in [Ba], examples of which appeared earlier in Figures 9 and 10 of Section 1 (see page 51). The contractions involved depend continuously on a rotation angle which is being increased gradually from 90° to 135°, and the sets below are thus points along the resulting path in $\mathcal{K}X$.

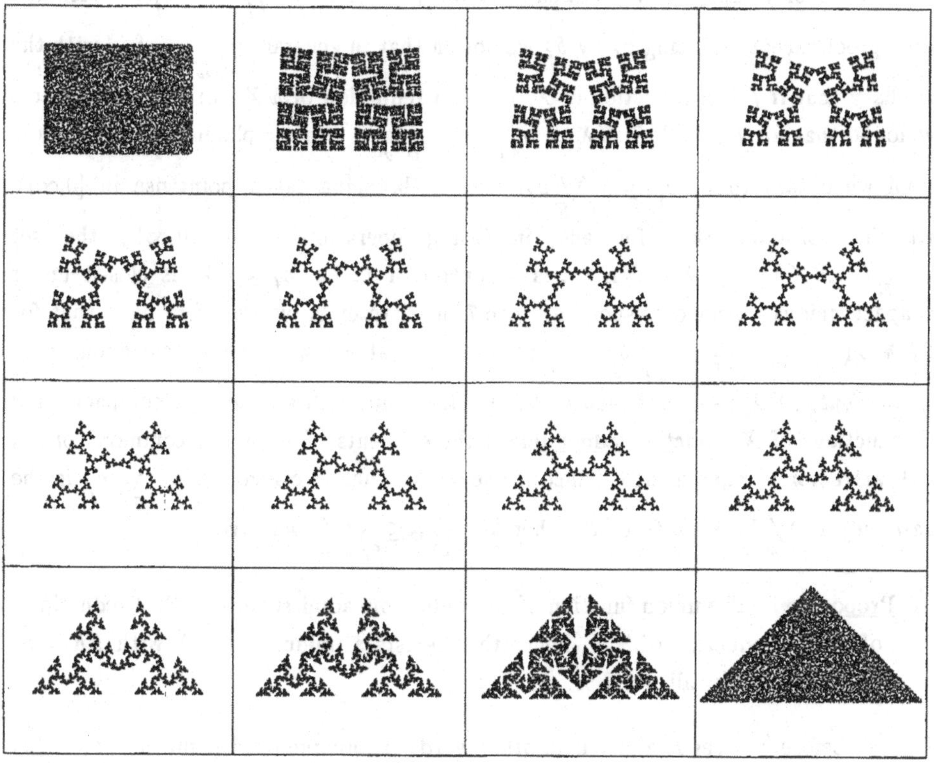

9. The Generalization from Contractions to Reductions

In Appendix 4 the notion of a 'reduction' f of X is described, generalizing the the notion of a contraction whilst retaining some of the properties; in particular, every reduction f has a metrically attractive fixed point (our space X being complete) and $\langle \operatorname{diam} f^n A \rangle$ strictly decreases to 0 for all $A \in \mathscr{K}X$. As might be expected then, the past theory dealing with compact sets F of contractions can largely be generalized to compact sets of reductions. In this section the necessary modifications will be described. Roughly speaking, the Lipschitz ratio r_f of a function f must be replaced by the modulus of continuity δ_f of f (see Appendix 4).

In 5.3 of Chapter 2 we showed that for $f \in C(X,Y)$, $r_{(f_{\mathscr{K}})} = r_f$. With the same proof except replacing r_f by δ_f we obtain that in general $\delta_{(f_{\mathscr{K}})} = \delta_f$, with the corollary that if f is a reduction of X, $f_{\mathscr{K}}$ is a reduction of $\mathscr{K}X$. In 6.5 of Chapter 2 we found that for $F \in \mathscr{K}F(Z, \mathscr{K}X)$, $r_{\bigcup F} \leq \bigvee_{f \in F} r_f$. Again replacing r_f by δ_f in the proof we obtain that $\delta_{\bigcup F} \leq \bigvee_{f \in F} \delta_f$ (the l.u.b. being taken pointwise in $[0,\infty]$), with the corollary that for any increasing operation u on $[0,\infty)$, the set $Cont_u(Z, \mathscr{K}X) = \{f\colon Z \to \mathscr{K}X \mid u \text{ is a control of } f, \text{ i.e. } \delta_f \leq u\}$ is closed under compact unions. And corresponding to 6.7 of Chapter 2, we therefore have that for $F \in \mathscr{K}C(X,X)$, $\delta_{\bigcup F} \leq \bigvee_{f \in F} \delta_f$, so for any u as above we have a continuous map $\bigcup : \mathscr{K}Cont_u(X,X) \to Cont_u(\mathscr{K}X, \mathscr{K}X)$. If we now call a nonempty compact set F of reductions of X <u>reduction-admissible</u> if the elements of F have a common control with attractor 0, equivalently (since if they have such a control then $\bigvee_{f \in F} \delta_f$ is the *least* one) if $\bigvee_{f \in F} \delta_f$ has attractor 0, then as $\delta_{\bigcup F} \leq \bigvee_{f \in F} \delta_f$ we have . . .

<u>9.1</u> <u>Proposition</u> The union function of any reduction-admissible set F of reductions of X is a reduction of $\mathscr{K}X$, hence there exists a unique $K \in \mathscr{K}X$ with $FK = K$. Furthermore, for all $A \in \mathscr{K}X$, $\langle F^n A \rangle \to K$. ○

As before, the set K above is in other words the unique fixed point of $\bigcup F$, and is called the <u>attractor</u> of F, denoted more fully by K_F, though we'll usually just write K. The term 'reduction-admissible' is of course the analogue of 'admissible' used in Section 1 (which could now be thought of as 'contraction-admissible'), and every admissible set of contractions is also reduction-admissible. And just as every nonempty *finite* set of contractions is admissible, we find that . . .

<u>9.2</u> <u>Proposition</u> Every nonempty finite set of reductions of X is reduction-admissible.

Proof:

Let F be such a set, and let $\delta = \bigvee_{f \in F} \delta_f$, which is an increasing operation on $[0,\infty)$ as each δ_f is . A4.2(1) gives a criterion for such an operation to have attractor 0 ; and since each δ_f obeys the condition on the right of that result, it's easily seen by finiteness of F that δ does too, so δ has attractor 0, i.e. F is reduction-admissible. o

In the case of *compact X*, the situation is particularly tidy ;

9.3 Proposition If X is compact, *every* nonempty compact set of reductions of X is reduction-admissible.

Proof:

Let $F \in \mathcal{K}\,Reduc\,X$ and $\delta = \bigvee_{f \in F} \delta_f$, which is increasing as each δ_f is. We prove that

(1) $\{\delta_f \mid f \in F\}$ is compact in the uniform topology.

(2) $\forall \epsilon \geq 0,\ \delta\epsilon = \max\{\delta_f\epsilon \mid f \in F\}$.

(3) $\forall \epsilon > 0,\ \delta\epsilon < \epsilon$.

(4) δ is lower continuous.

(5) δ has attractor 0, i.e. F is reduction-admissible.

For *any X*, the map sending each controlled operation f on X to its least control δ_f is continuous with respect to the uniform topologies (*Proof*: If $g \in \mu f$ then $\forall x,y \in X^*$ $g\,x \approx f\,x$ and $g\,y \approx f\,y$ so $d(g\,x,g\,y) \approx d(f\,x,f\,y)$, so by a nonstandard result on l.u.b.s, $\forall \epsilon \in [0,\infty)^*$, $\bigvee\{d(g\,x,g\,y) \mid d(x,y) \leq \epsilon\} \approx \bigvee\{d(g\,x,g\,y) \mid d(x,y) \leq \epsilon\}$, i.e. $\delta_g\epsilon \approx \delta_f\epsilon$. So $\delta_g \approx \delta_f$). As X is compact, the uniform topology on the domain of this map is the compact-uniform topology, so as F is compact with respect to it, (1) holds. $\forall \epsilon \geq 0$, since the map $\delta_f \mapsto \delta_f\epsilon$ is continuous it follows that $\{\delta_f\epsilon \mid f \in F\}$ is compact, hence has a greatest element which is thus $\delta\epsilon$, giving (2), and (3) follows. By (1) and the fact that each δ_f is lower continuous (see A4.1), (4) holds (*Proof*: Take any real $\epsilon \geq 0$ and $\gamma \approx \epsilon$ with $\gamma \geq \epsilon$. We must show that $\delta\gamma \approx \delta\epsilon$, i.e. (as $\delta\epsilon \leq \delta\gamma$) ${}^\circ\delta\gamma \leq \delta\epsilon$. Let the greatest element of $\{\delta_f\gamma \mid f \in F^*\}$ be $\delta_g\gamma$ and the greatest element of $\{\delta_f\epsilon \mid f \in F\}$ be $\delta_h\epsilon$. So $\delta\gamma = \delta_g\gamma$ and $\delta\epsilon = \delta_h\epsilon$, hence we must show that ${}^\circ\delta_g\gamma \leq \delta_h\epsilon$. Letting $f \in F$ with $\delta_g \in \mu\,\delta_f$ we have $\delta_g\gamma \approx \delta_f\gamma \approx \delta_f\epsilon \leq \delta_h\epsilon$ so ${}^\circ\delta_g\gamma \leq \delta_h\epsilon$ as required). Finally, (3) and (4) give (5) by A4.2(2). o

Note incidentally that replacing X by $\mathcal{K}X$ (and recalling that $\mathcal{K}X$ is compact iff X is compact), the above result shows that for compact X, $Reduc\,\mathcal{K}X$ is closed under compact unions.

The set of reduction-admissible sets F is denoted by ***Reduc–Admis X***. From now on assume $F \in$ *Reduc-Admis X* with attractor K, and let $\delta = \bigvee_{f \in F} \delta_f$. The necessary changes or modifications to the results in Sections 1 through 8 are now listed . . .

Section 1 (1.3 onwards). Replace *Admis X* by *Reduc-Admis X* in 1.3, 1.4 and 1.5, and replace r_f by δ_f in the proof of 1.3.

Section 2. Only a few obvious modifications. This time, for each nonempty $f \in F^{<\omega}$, $\circ f$ is a reduction of X with $\delta_{\circ f} \leq \delta^{|f|}$ so in general $h(fA, fB) \leq \delta^{|f|} h(A,B)$ and $\operatorname{diam} fA \leq \delta^{|f|} \operatorname{diam} A$.

Section 3. In the second paragraph and in the proof of 3.1, r should be replaced by δ. A similarly obvious change is made in the proofs of 3.7 and 3.9.

Section 4. In the proof of 4.5, replace the use of A3.4 by use of the corresponding result for reductions, namely A4.10.

Section 5. Replace r by δ in the proof of 5.5.

Section 6. No change.

Section 7. No change.

Section 8. Replace *Admis X* by *Reduc-Admis X*, so the attractor map under consideration is K: *Reduc-Admis* $X \to \mathcal{K}X$. In the proof of 8.1, the first map is now $\bigcup$: *Reduc-Admis* $X \to$ *Reduc* $\mathcal{K}X$ and second is fix : *Reduc* $\mathcal{K}X \to \mathcal{K}X$, which is continuous by A4.8 (applied to $\mathcal{K}X$). In 8.2 $Admis_u X$ should be replaced by ***Reduc-Admis***$_u X = \{F \in$ *Reduc-Admis* $X \mid \forall f \in F \ \delta_f \leq u$, i.e. $\bigvee_{f \in F} \delta_f \leq u\}$ for an increasing operation u on $[0,\infty)$ with attractor 0, and the modifications to the proof are obvious (the continuity of fix : $Reduc_u \mathcal{K}X \to \mathcal{K}X$ by A4.10 applied to $\mathcal{K}X$). In 8.3 *Contrac X* and $Contrac_u X$ should be replaced by *Reduc X* and $Reduc_u X$, for u as above.

10. Notes, Questions, and Suggestions for Further Work

In this section we list an assortment of notes, questions, and suggestions for further work, grouped under the sections to which they seem most appropriate. Assume unless otherwise implied that F is as usual an admissible set of contractions of a nonempty complete metric space X, with attractor K.

Section 1 : Existence of Invariant Sets

10.1 Let *Contrac-Sim* $\mathbb{R}^n$ be the set of contractive similitudes (by which I mean similitudes which are contractions) of $\mathbb{R}^n$. Then since the scale factor map r: *Contrac-Sim* $\mathbb{R}^n \to (0,\infty)$ is continuous, *every* nonempty compact set of contractive similitudes is admissible.

10.2 There exists compact X and $F \in \mathcal{K}\,\mathit{Contrac}\,X$ with $\bigcup F$ not a contraction (however, by 9.3 $\bigcup F$ is a *reduction* at least). Here is an example. Let $X = [0,1]$ and for $n \geq 2$ let $f_n : X \to X$ be as follows, where $\epsilon_n = 1/n$. On $[0, \epsilon_n]$ f_n contracts about 0 by factor $1-\epsilon_n$. On $[\epsilon_n, 1]$ f_n first contracts about ϵ_n by factor 1/2 then subtracts $f_n \epsilon_n$. Note that

(1) f_n is a contraction with ratio $1-\epsilon_n$.
(2) f_2 contracts about 0 by factor 1/2.
(3) $\langle f_n \rangle_{n \geq 2} \to f_2$.

So, letting $F = \{f_n \mid n \geq 2\}$, F is compact by (3). However, for all $n \geq 2$, $\epsilon_{n+1} < \epsilon_n$ so $(1-\epsilon_n)\epsilon_{n+1} = f_n \epsilon_{n+1} \in F\{\epsilon_{n+1}\}$ so $h(F\{\epsilon_{n+1}\}, F\{0\}) = h(F\{\epsilon_{n+1}\}, \{0\}) \geq (1-\epsilon_n)\epsilon_{n+1} = (1-\epsilon_n)\, h(\{\epsilon_{n+1}\}, \{0\})$ giving $r_{\bigcup F} \geq 1-\epsilon_n$. Thus $r_{\bigcup F} \geq 1$, so $\bigcup F$ is not a contraction.

10.3 For $F \in \mathcal{K}\,\mathit{Contrac}\,X$, we know that if F is admissible then $\bigcup F$ is a contraction; the converse however does *not* hold. For example, take $X = [0,1]$ and let $F = \{f_n \mid n \geq 2\} \cup \{c_x \mid x \in X\}$ where f_n is as described in the last note and c_x is the constant map with value x. Then $\bigcup F$ is the constant map with value X, but $\bigvee_{f \in F} r_f = 1$ so F is not admissible. Is it true though that for $F \subseteq \mathcal{K}\,\mathit{Reduc}\,X$ such that $\bigcup F$ is a reduction, F is reduction-admissible? If so, we could shorten the proof of 9.3 since, using A4.5 applied to $\mathcal{K}X$, 9.3(3) implies $\bigcup F$ is a reduction.

10.4 It's not generally true that F is determined by its union function $\bigcup F$; we can have $F \neq G$ yet $\bigcup F = \bigcup G$. For example let $X = \mathbb{R}$, $F = \{f_1, f_2\}$ and $G = \{g_1, g_2\}$ where $f_1 x = x/2$, $f_2 x = -x/2$, g_1 agrees with f_2 on $(-\infty,0]$ and with f_1 on $[0,\infty)$, and g_2 agrees with f_1 on $(-\infty,0]$ and with f_2 on $[0,\infty)$. To see that $\bigcup F = \bigcup G$ use the easy general result that $\bigcup F = \bigcup G \Leftrightarrow \forall x \in X\ Fx = Gx$ (Fx and Gx denoting $F\{x\}$ and $G\{x\}$). However, the maps g_i are not bijective. Is there an example in which all the elements of $F \cup G$ are bijective? Note also that in general $\bigcup F = \bigcup G \nRightarrow \{\text{fix } f \mid f \in F\} = \{\text{fix } g \mid g \in G\}$.

10.5 For any similitude g of $X = \mathbb{R}^n$, $g(K_F) = K_{(g \circ F \circ g^{-1})}$ where $g \circ F \circ g^{-1}$ denotes $\{g \circ f \circ g^{-1} \mid f \in F\}$.

Section 2 : nth-level Images

10.6 Suppose all elements of F are bijective. Recall that for $f \in F^n$ we defined $\circ f = f_0 \circ \cdots \circ f_{n-1}$, the terms of f being applied in *reverse* order of their appearance in f. It may be helpful to realise that $\circ f$ can also be expressed as a composition of suitable *conjugates* of $f_0, \ldots, f_{n-1}$ in this order. Namely, letting g^h denote the conjugate $h \circ g \circ h^{-1}$ of g by h, and using that $h \circ g = g^h \circ h$, we have for example that $f_0 \circ f_1 \circ f_2 \circ f_3 = (f_3^{f_0 \circ f_1 \circ f_2}) \circ (f_2^{f_0 \circ f_1}) \circ (f_1^{f_0}) \circ f_0$. The pattern is obvious.

10.7 The following result essentially appears in [Ha] on page 389; if all $f \in F$ are injective and $|K| \geq 2$ then K is perfect. A more basic proof is to use that for $x \in K$, for all n x belongs to some nth-level image fK of K, and the diameters of the nth-level images decrease to 0, and every nth-level image of K has the same cardinality as K due to injectivity of the elements of F.

Section 5 : Connectedness of K

10.8 Note that we can have $A \in \mathcal{K}X$ reduced by F with $\langle fA \mid f \in F \rangle$ interlinked yet $\langle fA \mid f \in F^2 \rangle$ *not* interlinked. An example is shown below in which $X = \mathbb{R}^2$ and $F = \{f_1, f_2\}$, where f_i contracts about x_i by factor 1/2 then rotates about p_i by $\pi/2$. In this example K_F is totally disconnected since F reduces FA and $\langle fFA \mid f \in F \rangle$ is disjoint.

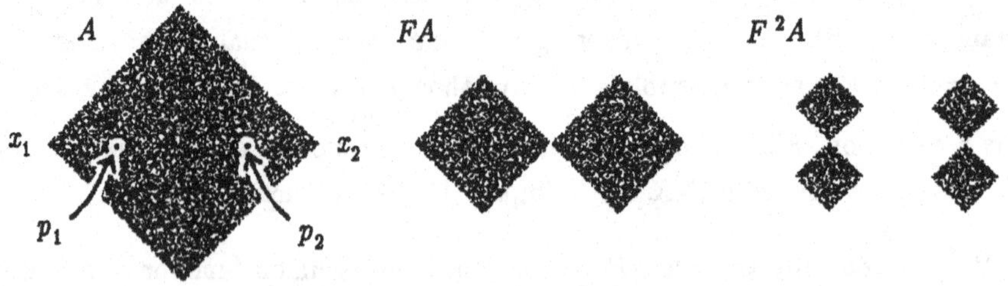

10.9 If every element of F is a homeomorphism and F satisfies the compact set condition with A such that $\forall f \in F \; fA \subseteq \underline{A}$, then FA is reduced by F with $\langle fFA \mid f \in F \rangle$ disjoint, so k is injective, hence K is totally disconnected if $|F| \geq 2$.

Section 6 : Regularity and Residuality of K

10.10 Try and find practical sufficient conditions for K to be regular.

Section 7 : Tiling of K

10.11 If F is finite and every element of F is a homeomorphism, and $\langle fK \mid f \in F\rangle$ is nonoverlapping in K, must each fK be regular in K (so K is tiled with respect to F)? A more general question is suggested; if a compact Hausdorff space K is the union of a finite nonoverlapping family $\langle K_i \mid i \in I\rangle$ of homeomorphs of K, must each K_i be regular in K? The answer to *this* however is no (unless $|F| = 2$, in which case the answer is yes by A5.3). For example, let K be the subspace $[0,1]^2 \cup \{(2,0)\}$ of $\mathbb{R}^2$, K_1 be $([0,1/2] \times [1/2,1]) \cup \{(3/4,1/2)\}$, K_2 be $[1/2,1]^2 \cup \{(1/4,1/2)\}$, and K_3 be $([0,1] \times [0,1/2]) \cup \{(2,0)\}$; then neither K_1 nor K_2 are regular in K.

Back to the *original* question, the answer is yes for *regular* K by 7.12(1), so we need only consider the case where K is residual in X. In search of an answer in the *negative*, a suitable example *may* lie in the following. Let $X = \mathbb{R}$, $\lambda \in (0,1/2)$, and $F_\lambda = \{f_1, f_2, f_3\}$ where f_1 contracts about 0 by factor λ, f_2 contracts about 0 by factor 1/2, and f_3 contracts about 1 by factor 1/4 say. Let K_λ denote the attractor of F_λ. Then is there λ such that $\langle fK_\lambda \mid f \in F_\lambda\rangle$ is nonoverlapping in K_λ yet some fK_λ is not regular in K_λ? For this, λ must not be a power of 1/2 or else f_1 will be a power of f_2 and we'll have $K_\lambda = K_{\{f_2, f_3\}}$, and f_1K will overlap f_2K in K.

Miscellaneous

10.12 Note that any convex $A \in \mathcal{K}\,\mathbb{R}^n$ is of the form K_F for some compact set F of contractive similitudes. For example we can take $F = \{f_a \mid a \in A\}$ where, for any $\lambda \in (0,1)$, each f_a is contraction about a by factor λ.

10.13 For a set S of similitudes of $\mathbb{R}^n$, try and determine characteristics of sets which are of the form K_F for some finite set $F \subseteq S$. In particular consider the case where S is the set of dilations.

10.14 The 'Collage Theorem' (see [Ba]), an instance of a result applying to *all* contractions, gives that $\forall A \in \mathcal{K}X \;\; h(A,K) \le h(A,FA)/(1-r)$ where $r = \bigvee_{f\in F} r_f$, which in particular gives $K \subseteq [A]_{h(A,FA)/(1-r)}$. Putting $A = \{\text{fix}\, f \mid f \in F\}$ and noting that $h(A,FA) \le r\,\text{diam}\,A$, we obtain $K \subseteq [A]_{r\,\text{diam}\,A\,/(1-r)}$. Or, putting $A = \{x\}$ for any $x \in X$, and noting that in this case $h(A,FA) = \max\{d(x,fx) \mid f \in F\}$, we have $K \subseteq [x]_{\max\{d(x,fx) \mid f\in F\}/(1-r)}$. The latter could be used in a computer program drawing a picture of K to automatically establish a region in which K lies, the program then making sure that the region is represented on the screen. One obvious choice for x would be to take x close to the fixed point of some element of F.

10.15 Throughout we've been using the compact-uniform topology on *Contrac* X, but we could instead have used the bounded-uniform topology, with hardly any changes to the work. On the positive side, using the latter topology we'd no longer need local compactness of X for continuity of the attractor map $F \mapsto K_F$. On the negative side, generally fewer sets $F \subseteq$ *Contrac* X would be admissible (i.e. nonempty and compact, with $\bigvee_{f \in F} r_f < 1$), since the latter topology is a refinement of the former. However, if X is boundedly compact, such as $\mathbb{R}^n$ which is the main case of interest, the two topologies coincide anyway.

10.16 Consider the possible generalization of Hutchinsons's invariant *measures* (see [Hu]) to the case of compact F. Likewise consider generalizing the random iteration algorithm (see [Ba]) to this more general case. Instead of probabilities $p_1, \ldots, p_n$ in the case of *finite* $F = \{f_1, \ldots, f_n\}$, we'd use a Borel probability measure on the topological space F.

10.17 There is a continuous analogue of Hutchinson's theory, which suggests itself when we recast the framework of the latter in terms of semiflows. In Hutchinson's theory we started by taking a nonempty compact set of contractions of X and forming a contraction of $\mathcal{K}X$; equivalently we started with a nonempty compact set of contractive *discrete semiflows* in X and formed a contractive discrete semiflow in $\mathcal{K}X$. The idea now is to try and replace *discrete* by *continuous*. First here are the answers to some pertinent questions.

(1) What is a discrete or continuous semiflow?

A *semiflow* in a space X is a topological monoid action θ of $(\mathbb{N},+)$ or $([0,\infty),+)$ on X, the semiflow being *discrete* or *continuous* according to whether the 'time set' T involved is $\mathbb{N}$ or $[0,\infty)$. We'll usually denote $(\theta t)x$ by x_t with θ in mind. The discrete semiflows correspond bijectively with the continuous operations f on X, f generating the discrete semiflow in which the action of $t \in \mathbb{N}$ is f^t.

(2) What topology do we put on the set $Act(T,X)$ of actions of T on X (where T is in general any locally compact Hausdorff monoid and X is a locally compact Hausdorff space)?

Each action of T on X is a continuous map $T \to C(X,X)$, and we'll therefore give $Act(T,X)$ the compact-open topology inherited from $C(T, C(X,X))$. We can describe this another way. For $\theta \in Act(T,X)$, let *eval* θ denote the evaluation map $T \times X \to X$ corresponding to θ. Each such map belongs to $C(T \times X, X)$, which we naturally give the compact-open topology. Then it is easily verified that the earlier topology on $Act(T,X)$ is the one induced by the map $\theta \mapsto eval\,\theta$. Thus, for $\theta \in Act(T,X)$ and $\varphi \in Act(T,X)^*$,

$$\varphi \in \mu\theta \Leftrightarrow \forall (t,x) \in T \times X \;\; \forall (s,a) \in \mu(t,x) \;\; (\varphi s)a \in \mu(\theta t)x .$$

In the case $T = (\mathbb{N},+)$, $Act(T,X) \cong C(X,X)$ via the correspondence noted in (1).

(3) What is a *contractive* semiflow?

Define the *ratio* r_θ of a semiflow θ in X to be the least $r \in [0,\infty]$ such that $\forall t \in T$ $\forall x,y \in X$ $d(x_t, y_t) \le r^t d(x,y)$, i.e. $\forall t \in T$ $r_{\theta t} \le r^t$. Then θ is *contractive* if $r_\theta < 1$. Note that if $T = \mathbb{N}$ with θ generated by f, $r_\theta = r_f$, so θ is contractive iff f is a contraction.

(4) Is there an analogue of Banach's contraction mapping theorem for continuous semiflows?

Yes. Every contractive continuous semiflow θ in a nonempty complete metric space X has a metrically attractive fixed point, i.e. a point p such that $\forall t$ $p_t = p$ and $\forall x \in X$ $d(x_t, p)$ strictly decreases to 0 as $t \to \infty$. This p is necessarily the unique fixed point of each θt for $t > 0$. The result can be proved nonstandardly using transfer of Banach's theorem as follows. Take infinite $n \in \mathbb{N}^*$, let $\tau = 1/n$ and let $f = \theta\tau$. Then f is a $*$-contraction, so let its fixed point be p say. As $f^n = \theta 1$, $p = \text{fix } f^n = \text{fix } \theta 1 \in X$. p is fixed under θ since $\forall t > 0$, where $m\tau \in \mu t$ then $p = f^m p = p_{m\tau} \in \mu\, p_t$ so $p = p_t$. Lastly, $\forall x \in X$, $d(x_t, p)$ strictly decreases to 0 as $t \to \infty$, using that $d(x_t, p) = d(x_t, p_t)$ $\le r_\theta^{\,t}\, d(x,p)$ and, if $x_t \ne p$ and $s > 0$, $d(x_{t+s}, p) \le r_\theta^{\,s}\, d(x_t, p) < d(x_t, p)$.

Assume now that our metric space X is locally compact, and suppose Θ is a nonempty compact set of contractive continuous semiflows in X (with respect to the topology described in (2)), with $\bigvee_{\theta \in \Theta} r_\theta < 1$. We now sketch the nonstandard method of obtaining a continuous semiflow in $\mathcal{K}X$ (the 'union semiflow' of Θ). Taking infinitesimal $\tau > 0$, let $F = \{\theta\tau \mid \theta \in \Theta^*\}$, which belongs to $(Admis\ X)^*$, and has invariant set K say, the fixed point of the $*$-contraction $\bigcup F$ of $\mathcal{K}X^*$. Then if all is well (note; we may need to impose a further 'taming' condition on Θ for this, and perhaps also assume X is *boundedly* compact; the details have not yet been fully checked), we should then be able to obtain a continuous semiflow ψ in $\mathcal{K}X$ such that for $t \in [0,\infty)$ and $n\tau \in \mu t$, $(\bigcup F)^n \in \mu\, \psi t$; intuitively $\bigcup F$ serves as an 'infinitesimal generator' for the semiflow. And this semiflow will be contractive, with fixed point $st\, K$ (the 'attractor' or 'invariant set' of Θ, as it would be called). As mentioned above, some details remain to be checked; but the work should go through. As a simple example, if Θ consists of three continuous semiflows contracting steadily about the vertices of a triangle, the invariant set will be the triangle in question.

Chapter 4

Views and Fractal Notions

0. Introduction

This chapter finds its origins in what was perceived to be something of an imbalance in the study of 'fractals', a term which in the absence of any universally agreed definition is being used here informally to refer to sets having 'detail beyond detail' in some sense, perhaps also having some form of self-similarity. Namely, whilst a considerable amount of work seems to be under way on the dimension-theoretic side (studies involving Hausdorff dimension for example), there seems to be little being done on the more *visual* side, and the present chapter represents an attempt to begin redressing the balance. Specifically, the simple idea of a 'view' is introduced as the basic constituent in a framework designed for use in studying the structure of sets within a given space X. Particularly in the case of $X = \mathbb{R}^2$, the type of structure studied can be thought of as *visual* structure, as the term 'view' already suggests. Imagining looking down on a subset A of the plane $X = \mathbb{R}^2$, what we'd see would be a region D of X along with the part $A \cap D$ of A lying *within* that region. Such a pair $(D, A \cap D)$ is an example of what we'll formally be defining as a 'view of A', D being the 'domain' of the view, and it should be apparent that this definition can be generalized not only to subsets A of *any* topological space X but, in *fullest* generality, to subsets A of arbitrary *sets* X. This then is the basic idea of views, which can be used to express a variety of concepts to investigate. Indeed, the formal notion of a view was originally designed in *pursuit* of formulating one of these concepts, namely a certain type of self-similarity of closed subsets A of $\mathbb{R}^2$. The rough idea was that wherever one looked at A one should see the same sort of structure. Or, rephrased somewhat more definitely, that every nonempty view of A should be embedded in every other nonempty view of A, in a sense suggested by the picture below;

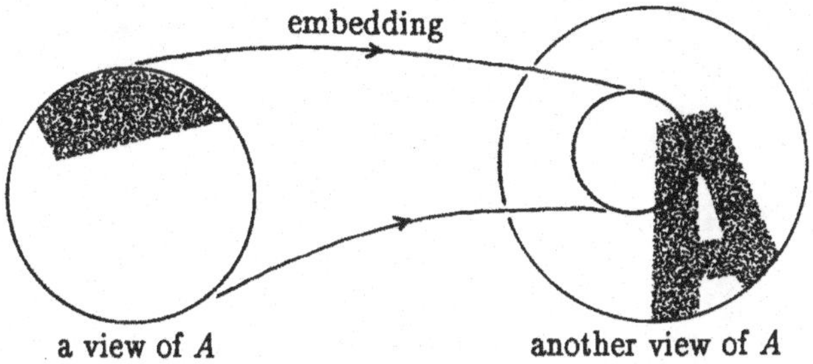

This definition of embedding implicitly involves a group of transformations of X, namely the group of direct similitudes, which also induces an obvious and closely related notion of when two views are similar (namely when some direct similitude maps the one view to the other). However, in a like way *any* group of transformations of X gives rise to a notion of similarity and embeddings of views so it will be natural to generalize the framework accordingly. In the general setting we'd abstractly refer to the elements of the group G involved as 'similarities', modelling things on the primary example in which $X = \mathbb{R}^n$ and the similarity group G consists of the direct similitudes. Another natural similarity group to consider in the case of $\mathbb{R}^n$ would be the group of affine bijections. Most of the view-related definitions we'll introduce, particularly that of a 'view class', will involve the presence of a similarity group.

We now outline the work to follow. In Section 1 we give in full generality the basic definitions concerned with views and similarities, with remarks towards the end on the case of topological spaces which will form the setting of subsequent applications. Section 2 is concerned with what might be thought of as the relative 'visual power' of the view structures described in Section 1, and also of the augmented structures in which a similarity group is present. In Section 3 we formally introduce the above-mentioned notion of self-similarity along with some related concepts, proving a few properties of self-similar closed subsets of a Hausdorff space, and in Section 4 we show that certain 'ω-extensions' of Hutchinson's invariant sets are self-similar. In Section 5 we study three topologies arising in the context of views of closed subsets of a Hausdorff space; foremost is the 'view topology' on the set of views, and this induces a topology on the closed sets and a topology on the 'view classes'. Section 6 uses the view topology in defining when a subset A of $\mathbb{R}^n$ is 'visually fractal' at a point $x \in A$, namely expressing that as we zoom in on x, what we see never settles down, so that in this sense there is detail beyond detail. If this holds at all points of A we say that A is 'visually fractal', and it is subsequently shown that many invariant sets have this property. Finally we have Section 7, providing a list of notes, questions, and suggestions for further work.

1. Views and Similarities

In this section we give in full generality the basic definitions concerned with views of subsets of a set X. We start with the framework in which views are defined . .

A view space is a pair $(X, \mathscr{D})$ where X is a set and $\mathscr{D}$ is a view structure on X, namely a set of nonempty subsets of X. X is called the domain of the space and $\mathscr{D}$ the view structure of the space, its elements being called the view domains , thought of as the 'observable bubbles of space', more accurately as the regions of X one can see at a single glance. The primary examples we have in mind are where X is $\mathbb{R}^n$ and $\mathscr{D}$ is the set of open balls (the usual view structure on $\mathbb{R}^n$ as we'll call it), in particular the case of $\mathbb{R}^2$, which the pictures in this and later sections illustrate. A key point here is that the view domains are all *bounded*, although they also cover arbitrarily large regions ; such a view structure represents the property of being able to see only *bounded* regions at a glance, albeit arbitrarily large ones, which can be considered an idealization of reality. Returning to the general situation, with $(X, \mathscr{D})$ in mind along with a subset ***Ob*** (the object set) of $\mathscr{P}X$ whose elements we shall call objects (thought of as the 'objects of study') we make the following definitions concerned with 'views' of objects.

For $D \in \mathscr{D}$ and $A \in Ob$, the D-view of A is $DA = (D, A \cap D)$. An entity v of this form is called a view, this particular one being a view of A. D is the domain of v , denoted by ***dom v***, and $A \cap D$ is the object part of v , denoted by ***ob v***, thought of as the part of an object (which you might like to think of as being coloured black in an otherwise white space) visible in D. If the object part of v is *all* of A we may say v is a whole view of A.

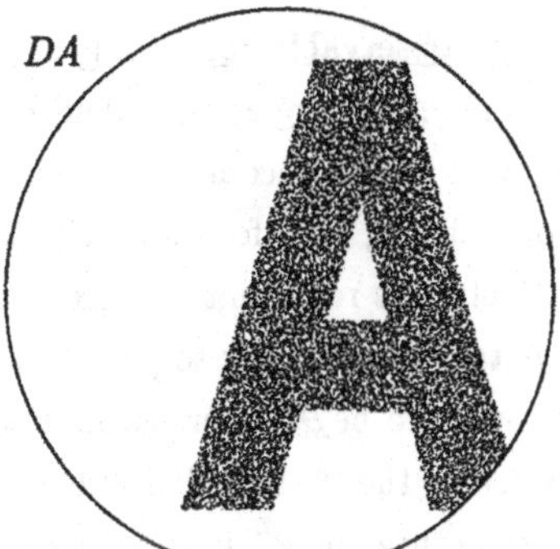

The set of views will normally be denoted by $\mathscr{V}$. However, for $\mathscr{E} \subseteq \mathscr{D}$ and $\mathscr{A} \subseteq Ob$ we define $\mathscr{E}\mathscr{A} = \{EA \mid E \in \mathscr{E} \text{ and } A \in \mathscr{A}\}$, giving us the more explicit notation $\mathscr{D}Ob$ for the set of views, should it be desired. We generally abbreviate $\mathscr{E}\{A\}$ to $\mathscr{E}A$, so in particular the set of views of A is denoted by $\mathscr{D}A$. The set $\{A \cap D \mid A \in Ob \text{ and } D \in \mathscr{D}\}$ of object parts will be denoted by ***Ob Parts***. In the following, the letters D and E will be assumed always to denote view domains whilst A, B, C will denote objects and u , v , w will denote views.

It is trivial but central to realise that an object A is not necessarily recoverable from a view of A since a view only reveals the part lying in its domain. This prompts the following terminology. We say an object A is consistent with a view v (or v is consistent with A) if v is a view of A, i.e. if $ob\ v = A \cap dom\ v$. This expresses that we *could* be looking at A. Generalising, we say A is **consistent** with a *set* $\mathcal{U}$ of views if it is consistent with every element of $\mathcal{U}$, i.e. $\mathcal{U} \subseteq \mathcal{D}A$. If there *exists* an object consistent with $\mathcal{U}$ we say $\mathcal{U}$ is **consistent**.

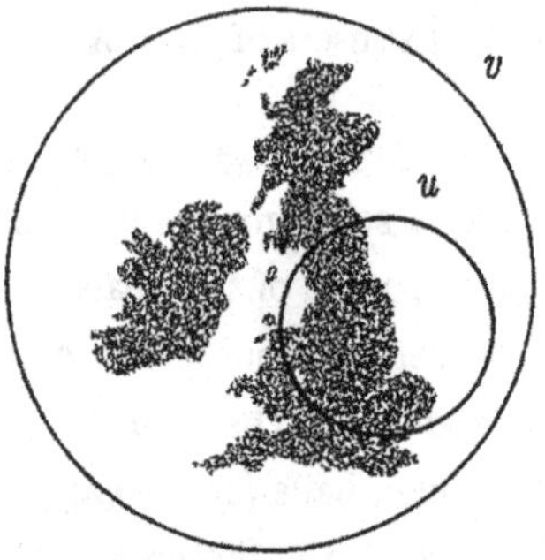

D is said to be a **sub view domain** of E if $D \subseteq E$. We say u is a **subview** of v, written $u \leq v$, if $dom\ u \subseteq dom\ v$ and $ob\ u = dom\ u \cap ob\ v$. This gives a partial ordering of $\mathcal{V}$, and synonymous to saying that u is a subview of v we may say that v is a **superview** of u.

We say v is **empty** if it has empty object part, expressing that no part of an object is visible in v. At the other extreme we shall say v is **full** if $ob\ v = dom\ v$, expressing that nothing *but* object is visible in v.

For $D \in \mathcal{D}$ we say objects A and B are **D-indistinguishable** if $DA = DB$, i.e. $A \cap D = B \cap D$. More generally, for $\mathcal{E} \subseteq \mathcal{D}$ we say A and B are **$\mathcal{E}$-indistinguishable** if A and B are D-indistinguishable for all $D \in \mathcal{E}$, equivalently if $A \cap \bigcup \mathcal{E} = B \cap \bigcup \mathcal{E}$. In the special case $\mathcal{E} = \mathcal{D}$ we may just say A and B are **view-indistinguishable**.

The **observable space** is $\bigcup \mathcal{D}$, the union of the view domains, and if this is all of X we say $\mathcal{D}$ is **covering**. Another important property relating to what might be thought of as the encompassing power of $\mathcal{D}$ is the following. We say $\mathcal{D}$ is **ideal** if it is an ideal basis, i.e. for all $D_1, D_2 \in \mathcal{D}$ there's $D_3 \in \mathcal{D}$ such that $D_1 \cup D_2 \subseteq D_3$. Intuitively this represents the property that given any two view domains, you can always take a step back to get a view domain encompassing both the former. A subset of X is said to be **$\mathcal{D}$-bounded** if it is covered by some element of $\mathcal{D}$. In the case where $\mathcal{D}$ is ideal, the $\mathcal{D}$-bounded subsets of X form the ideal generated by $\mathcal{D}$. The usual view structure on $\mathbb{R}^n$ is of course covering and ideal, and '$\mathcal{D}$-bounded' just means 'bounded'.

So far the framework described admits no notion of *similarity* between views and is consequently rather static in nature, suggesting little in the way of interesting concepts to investigate. It is the presence of a notion of similarity that will make things far more interesting. To this end we'll now assume we have a group G of permutations of X, under which $\mathcal{D}$ is closed. The triple $(X, G, \mathcal{D})$ will be called a **similarity view space**, $(G, \mathcal{D})$ being a **similarity view structure on X**. The elements of G (the **similarity group**) are called the **similarities**, denoted usually by f, g, h. The primary examples we have in mind are where X is $\mathbb{R}^n$, G is the group of direct similitudes (see Appendix 9) and $\mathcal{D}$ is the usual view structure on $\mathbb{R}^n$; $(G, \mathcal{D})$ will be referred to as the **usual similarity view structure on $\mathbb{R}^n$**. Returning to the general case, with $(X, G, \mathcal{D})$ in mind along with an object set Ob which is *closed* under G, (for example, the set of closed subsets of $\mathbb{R}^n$ in the examples just mentioned) we make the following definitions.

We say objects A and B are similar, written $A \sim B$, if B is the image of A under some similarity, in other words if B is in the orbit of A under the natural group action of G on Ob. The equivalence classes with respect to $\sim$ will be called object classes, the object class of A being denoted by $A^{\sim}$. Sometimes we may just say B is a copy of A if B is similar to A ; and if $g \in G$ maps A to B we may write $A \xrightarrow{g} B$.

Identical definitions to the above go for view domains ; D and E are similar, written $D \sim E$, if E is the image of D under some similarity, i.e. they're in the same orbit, and so on. This time the equivalence classes are called view domain classes. In addition we say D is embedded in E, written $D \rightarrow E$, if there is a similarity mapping D to a sub view domain of E, in other words if D is similar to a sub view domain of E. If g maps D into E we'll say g embeds D in E and write $D \xrightarrow{g} E$. Naturally, $D \xrightarrow{g} E \xrightarrow{h} F \Rightarrow D \xrightarrow{h \circ g} F$, and $\rightarrow$ is a preordering of $\mathscr{D}$.

Since $\mathscr{D}$ and Ob are both closed under G, so is the set of object parts, as $g(A \cap D) = gA \cap gD$. We therefore have a natural group action of G on $\mathscr{V}$ defined by $gv = (g\ dom\ v, g\ ob\ v)$, in other words comprising the actions of G on $\mathscr{D}$ and $Ob\ Parts$ working in parallel. The action is equivalently described by $g\,DA = gD\,gA$, since $g\,DA = g\,(D, A \cap D) = (gD, g\,(A \cap D)) = (gD, gA \cap gD) = gD\,gA$. The action of G on $\mathscr{V}$ gives us the definition of similarity between views ; u and v are similar, written $u \sim v$, if there's a similarity taking u to v, i.e. if they're in the same orbit. We may synonymously say v is a copy of u, and if g maps u to v we may write $u \xrightarrow{g} v$. The equivalence classes will be called view classes, the view class of u being denoted by $u^{\sim}$.

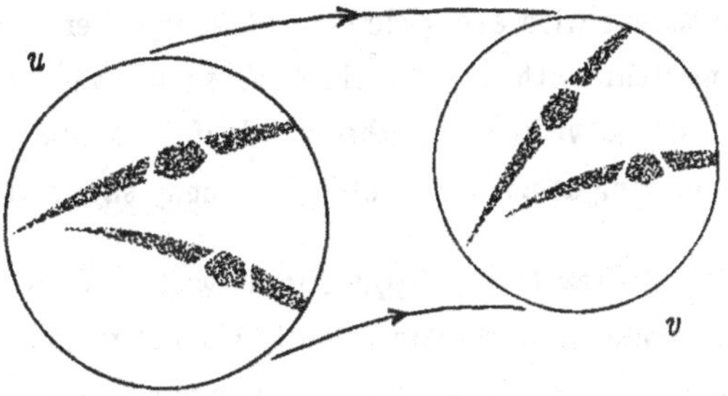

In terms of similarities we can now define the notion of one view being embedded in another as mentioned and illustrated in the introduction. Namely, we say u is embedded in v, written $u \rightarrow v$, if there is a similarity taking u to a subview of v, in other words if u is similar to a subview of v. If g maps u to a subview of v we may say g embeds u in v and write $u \xrightarrow{g} v$. Naturally, $u \xrightarrow{g} v \xrightarrow{h} w \Rightarrow u \xrightarrow{h \circ g} w$, and $\rightarrow$ preorders the set of views. Of course, if g embeds u in v it in particular embeds $dom\ u$ in $dom\ v$.

For a set $\mathscr{U}$ of views we define $\mathscr{U}^{\sim} = \{u^{\sim} \mid u \in \mathscr{U}\}$. In particular the set of view classes is thus denoted by $\mathscr{V}^{\sim}$, and the capitals U, V, W will be assumed to denote view classes in the following. The elements of a view class may suggestively be called its __realizations__. As with views we can think of view classes as being visual information about an object highlighted in black in an otherwise white space ; but whereas a view gives you *absolute* information in that you know exactly what domain you're looking at and what's visible in it, a view *class* only gives information modulo similarity. It's as if you're receiving an image which has been taken by a remote camera whose bearings are unknown ; more precisely the view domain the camera is looking at is only known modulo similarity. The transmitted information then is cloaked in a coat of uncertainty ; the original view v has become $v^{\sim}$, which *any* realization of $v^{\sim}$ could have given. This then is the significance of the concept of a view class, and informally you may like to think of them as 'views transmitted by remote camera'.

A number of definitions regarding views have natural counterparts for view classes. A __view class of A__ is a view class of a view of A ; the set of view classes of A is thus $(\mathscr{D}A)^{\sim}$. Now just as A is not necessarily recoverable from a view of A, neither of course is it necessarily recoverable from a view *class* of A, in fact even *less* so, and we may say A is __consistent__ with V (or vice versa) if V is a view class of A, intuitively expressing that we *could* be looking at A (with our remote camera). This definition formally conflicts with an earlier one in that V is also a set of views, and consistency with A in the above sense does not equate with consistency in the earlier sense of a set of views, but under the sensible assumption that the *above* definition is the one involved when dealing specifically with a view class, no confusion should arise. Generalizing the definition, A is __consistent__ with a *set* $\mathscr{U}$ of view classes if it is consistent with every element of $\mathscr{U}$, in other words if $\mathscr{U} \subseteq (\mathscr{D}A)^{\sim}$. Of course, if A is consistent with a view class V, so is every copy of A. If there *exists* an object consistent with $\mathscr{U}$, in other words if $\mathscr{U}$ is a bundle of information we *could* obtain were we looking at a suitable object, we may say $\mathscr{U}$ is __consistent__.

We say U is __embedded__ in V, written $U \rightarrow V$, if some element of U is embedded in some element of V, equivalently if *all* elements of U are embedded in *all* elements of V. Thus $u^{\sim} \rightarrow v^{\sim} \Leftrightarrow u \rightarrow v$. The view classes are preordered by $\rightarrow$.

A view class containing an empty view contains *only* empty views ; such a view class is said to be __empty__. So in general, $v^{\sim}$ is empty $\Leftrightarrow$ v is empty. Likewise, a view class is __full__ if its elements are full; equivalently, $v^{\sim}$ is full $\Leftrightarrow$ v is full.

The idea of view embedding leads naturally to two basic relations on the set of objects. We'll say a view u is __embedded__ in B, written $u \rightarrow B$, if u is similar to a view of B. Moreover if g maps u to a view of B we'll say g __embeds__ u in B, written $u \xrightarrow{g} B$. Note that g embeds DA in B iff $g\,DA = gD\,B$, equivalently $g\,(A \cap D) = B \cap gD$, i.e. $gA \cap gD = B \cap gD$; this will be frequently used in subsequent work. We now say A is __view-embedded__ in B, written $A \rightarrow B$, if every view of A is embedded in B, equivalently if every view class of A is a view class of B, i.e. $(\mathscr{D}A)^{\sim} \subseteq (\mathscr{D}B)^{\sim}$. This can be equivalently phrased as the fact that whenever A is consistent with a set of view classes, so is B; or in short, the possibility of A (i.e. the possibility that we're looking at A) implies the possibility of B. The relation $\rightarrow$ preorders the set of objects. We say A and B are **view-similar** (or **view class indistinguishable**), written $A \leftrightarrow B$, if $A \rightarrow B$ and $B \rightarrow A$, i.e. $(\mathscr{D}A)^{\sim} = (\mathscr{D}B)^{\sim}$, i.e. A and B have identical view classes, intuitively expressing that A and B cannot be distinguished by images from a remote camera. This is of course an equivalence relation, and it's easily seen that . . .

1.1 Note Similar objects are view-similar.

Proof:
Suppose $A \overset{g}{\sim} B$. Then $\forall D \in \mathscr{D}$, g embeds DA in B since $g\,DA = gD\,gA = gD\,B$. And since g gives a permutation of $\mathscr{D}$, we thus have $A \leftrightarrow B$. ∘

Significantly however the converse is false, even in the case of the usual similarity view structure on $\mathbb{R}^n$; see note 7.10. In the case where $\mathscr{D}$ is covering but *not* ideal, counterexamples are more easy to produce. For example, consider the modification to the usual similarity view structure on $\mathbb{R}^n$ in which the view domains are instead only the open balls with radius ≤ 1. Let A be a singleton and B consist of two points which are distance ≥ 2 apart. Then because all the view domains have diameter ≤ 2, A and B will be view similar; but they are not similar. More generally, let A be a nonempty closed set and B be the union of a disjoint family of copies A_i of A each produced from A by an *isometry*, such that points in distinct copies are always at least distance ≥ 2 apart. Then A and B will be view-similar, but they need not be similar of course.

So far we've introduced view structures and similarity view structures on a set X along with the main basic definitions concerning views of subsets of X. We conclude this section by describing the natural upgrading of the two types of structure to the *topological* setting. We can lead into this by considering a condition on a view structure $\mathscr{D}$ on a set X concerned with what could be thought of as the 'resolution power' of $\mathscr{D}$. Namely, if $\mathscr{D}$ covers X, and for all $x \in X$ and $D_1, D_2 \in \mathscr{D}$ containing x there's $D_3 \in \mathscr{D}$ with $x \in D_3 \subseteq D_1 \cap D_2$, we'll say $\mathscr{D}$ is **topological**. The reason for this is that the definition in other words says that $\mathscr{D}$ covers X and the intersection of any two elements of $\mathscr{D}$ is a union of elements of $\mathscr{D}$; i.e. $\mathscr{D}$ forms a basis for a topology on X. The idea of views has thus led us to the notion of a basis for a topology on a set, which in turn leads to topological notions of course. In a different world then, topology could conceivably have developed from the idea of views, but we shall not pursue this speculation here. Instead we now point out that topological view structures can equivalently be considered as the 'natural' view structures on topological spaces, as follows.

For a topological space $(X,\mathcal{O})$, a view structure on $(X,\mathcal{O})$ is a view structure on X which is a basis for $\mathcal{O}$ (i.e. it's topological and gives the topology of the space); in other words it's a basis for $\mathcal{O}$ not containing $\emptyset$. The pair $((X,\mathcal{O}),\mathcal{D})$ will be called a topological view space. Now where for a topological view structure $\mathcal{D}$ on a set X $\mathcal{O}_{\mathcal{D}}$ denotes the topology for which $\mathcal{D}$ is a basis , $\mathcal{D}$ is thus a view structure on $(X,\mathcal{O}_{\mathcal{D}})$. So to consider a topological view structure on a *set* is essentially to consider a view structure on a topological space. We'll favour the latter viewpoint since in practice we'll usually be starting off with a particular topological space in mind. All the definitions regarding view spaces naturally also go for topological view spaces (except that this time the domain of the space is defined as the topological space involved). Note by the way that for an *ideal* view structure $\mathcal{D}$ on a topological space, every nonempty compact set C is covered by a view domain, i.e. there's a whole view of C (*Proof*: $\mathcal{D}$ covers C so there's a finite subcover, and in turn some element of $\mathcal{D}$ covers all the elements of the finite subcover, hence covers C).

Regarding similarities, a similarity view structure on $(X,\mathcal{O})$ is a pair $(G,\mathcal{D})$ where G is a group of *homeomorphisms* of $(X,\mathcal{O})$ and $\mathcal{D}$ is a view structure on $(X,\mathcal{O})$ closed under G. The triple $((X,\mathcal{O}),G,\mathcal{D})$ is called a topological similarity view space, and all the definitions applying to similarity view spaces naturally apply here too. Note that the usual similarity view structure on the *set* $\mathbb{R}^n$ is moreover a similarity view structure on the *space* $\mathbb{R}^n$.

2. Relative Strength of View Structures and Similarity View Structures

Until further notice let X be a set. For view structures $\mathscr{D}_1$ and $\mathscr{D}_2$ on X we say $\mathscr{D}_1$ is **weaker than or equivalent to** $\mathscr{D}_2$, written $\mathscr{D}_1 \lesssim \mathscr{D}_2$, if

(1) Every element of $\mathscr{D}_1$ is covered by an element of $\mathscr{D}_2$, and

(2) Every element of $\mathscr{D}_1$ is a union of elements of $\mathscr{D}_2$.

We synonymously say that $\mathscr{D}_2$ is **stronger than or equivalent to** $\mathscr{D}_1$. Note that the relation $\lesssim$ is really the conjunction of the weaker relations given by (1) and (2), which respectively express that the 'encompassing power' and 'resolution power' of $\mathscr{D}_2$ is at least as good as that of $\mathscr{D}_1$ (noting that (2) can be reformulated as 'for all $x \in X$ and $D_1 \in \mathscr{D}_1$ with $x \in D_1$, there's $D_2 \in \mathscr{D}_2$ with $x \in D_2 \subseteq D_1$'). In short then, the definition expresses that $\mathscr{D}_2$ represents a power of vision at least as good as that of $\mathscr{D}_1$. In particular, if objects A and B are view-distinguishable with respect to $\mathscr{D}_1$ (i.e. for some $D \in \mathscr{D}_1$ $DA \neq DB$) they're also view-distinguishable with respect to $\mathscr{D}_2$, simply by condition (1). We say $\mathscr{D}_1$ and $\mathscr{D}_2$ are **equivalent**, written $\mathscr{D}_1 \sim \mathscr{D}_2$, if $\mathscr{D}_1 \lesssim \mathscr{D}_2$ and $\mathscr{D}_2 \lesssim \mathscr{D}_1$. The relation $\lesssim$ is a preordering of the set of view structures on X (since each of (1) and (2) define preorderings), so $\sim$ is an equivalence relation. Trivially, $\mathscr{D}_1 \subseteq \mathscr{D}_2 \Rightarrow \mathscr{D}_1 \lesssim \mathscr{D}_2$. Equivalent view structures are thought of as representing equal powers of vision. In particular note that if $\mathscr{D}_1 \sim \mathscr{D}_2$, objects A and B are view-distinguishable with respect to $\mathscr{D}_1$ iff they're view-distinguishable with respect to $\mathscr{D}_2$; or put in the contrapositive, they're view-indistinguishable with respect to $\mathscr{D}_1$ iff they're view-indistinguishable with respect to $\mathscr{D}_2$. View-indistinguishability is thus **equivalence invariant**; in general we apply this term to any view-related concept which remains unchanged if we to switch to an equivalent view structure.

2.1 Proposition For any view structure $\mathscr{D}$ on X there's a largest which is equivalent to $\mathscr{D}$, namely $M_{\mathscr{D}} = \{U \mid U$ is a nonempty union of elements of $\mathscr{D}$ covered by an element of $\mathscr{D}\}$.

Proof:

$\mathscr{D} \subseteq M_{\mathscr{D}}$ gives $\mathscr{D} \lesssim M_{\mathscr{D}}$, whilst by the definition of $M_{\mathscr{D}}$, $M_{\mathscr{D}} \lesssim \mathscr{D}$; so $M_{\mathscr{D}} \sim \mathscr{D}$. And for any view structure $\mathscr{E} \sim \mathscr{D}$, $\mathscr{E} \lesssim \mathscr{D}$ says $\mathscr{E} \subseteq M_{\mathscr{D}}$. ○

Calling $\mathscr{D}$ **maximal** if it equals $M_{\mathscr{D}}$, i.e. if it has no proper expansion to an equivalent view structure, then

2.2 Corollary A view structure $\mathscr{D}$ on X is maximal iff it is closed under $\mathscr{D}$-bounded unions, i.e. for any nonempty $\mathscr{E} \subseteq \mathscr{D}$ such that $\bigcup \mathscr{E}$ is $\mathscr{D}$-bounded, $\bigcup \mathscr{E} \in \mathscr{D}$. ○

Also note that $\mathscr{D}_1 \lesssim \mathscr{D}_2 \Leftrightarrow M_{\mathscr{D}_1} \subseteq M_{\mathscr{D}_2}$, so $\mathscr{D}_1 \sim \mathscr{D}_2 \Leftrightarrow M_{\mathscr{D}_1} = M_{\mathscr{D}_2}$. Restricted to the maximal view structures then, $\lesssim$ is the partial ordering $\subseteq$. More generally note that for maximal $\mathscr{D}_2$ and *any* $\mathscr{D}_1$, $\mathscr{D}_1 \lesssim \mathscr{D}_2 \Leftrightarrow \mathscr{D}_1 \subseteq \mathscr{D}_2$.

In the case where X is a metric space and $\mathscr{D}$ is the set of open balls, $M_{\mathscr{D}}$ is the set of nonempty bounded open sets. Of course, $M_{\mathscr{D}}$ is *topological* and gives the same topology as $\mathscr{D}$, namely that of X. This could have been anticipated by the following ;

2.3 Proposition For equivalent view structures $\mathscr{D}_1$ and $\mathscr{D}_2$ on X, if $\mathscr{D}_1$ is topological then so is $\mathscr{D}_2$, and $\mathscr{D}_2$ gives the same topology as $\mathscr{D}_1$.

Proof:

Firstly we show $\mathscr{D}_2$ is topological. Given any D_2, $E_2 \in \mathscr{D}_2$ we must show there is $F_2 \in \mathscr{D}_2$ with $x \in F_2 \subseteq D_2 \cap E_2$. As D_2 and E_2 are unions of elements of $\mathscr{D}_1$ let $D_1, E_1 \in \mathscr{D}_1$ with $x \in D_1 \subseteq D_2$ and $x \in E_1 \subseteq E_2$. Thus $x \in D_1 \cap E_1 \subseteq D_2 \cap E_2$, and as $\mathscr{D}_1$ is topological let $F_1 \in \mathscr{D}_1$ with $x \in F_1 \subseteq D_1 \cap E_1$. As F_1 is a union of elements of $\mathscr{D}_2$ let $F_2 \in \mathscr{D}_2$ with $x \in F_2 \subseteq F_1$. Then $x \in F_2 \subseteq D_2 \cap E_2$ as required.

Secondly, since each element of $\mathscr{D}_1$ is a union of elements of $\mathscr{D}_2$ and vice versa, $\mathscr{D}_1$ and $\mathscr{D}_2$ give the *same* topology. o

For any topological view structures $\mathscr{D}_1$ and $\mathscr{D}_2$ on X with $\mathscr{D}_1 \lesssim \mathscr{D}_2$, the topology given by $\mathscr{D}_2$ is a refinement of (i.e. is stronger than or equivalent to) that given by $\mathscr{D}_1$ of course. In short then, the stronger the topological view structure, the stronger the topology. Also, it's worth pointing out that . . .

2.4 Note For view structures $\mathscr{D}_1$ and $\mathscr{D}_2$ on a topological space $(X, \mathcal{O})$,

$$\mathscr{D}_1 \lesssim \mathscr{D}_2 \Leftrightarrow \text{Every element of } \mathscr{D}_1 \text{ is covered by an element of } \mathscr{D}_2.$$

Proof:

$\Leftarrow$: In addition we already know that every element of $\mathscr{D}_1$ is a union of elements of $\mathscr{D}_2$ because it belongs to $\mathcal{O}$ for which $\mathscr{D}_2$ is a basis. o

We now turn to the matter of comparing similarity view structures. For similarity view structures $(G_1, \mathscr{D}_1)$ and $(G_2, \mathscr{D}_2)$ on X we say $(G_1, \mathscr{D}_1)$ is **weaker than or equivalent to** $(G_2, \mathscr{D}_2)$, written $(G_1, \mathscr{D}_1) \lesssim (G_2, \mathscr{D}_2)$, if G_2 is a subgroup of G_1 and $\mathscr{D}_1 \lesssim \mathscr{D}_2$. Synonymously $(G_2, \mathscr{D}_2)$ is **stronger than or equivalent to** $(G_1, \mathscr{D}_1)$. The relation $G_2 \subseteq G_1$ implies that under the similarity group G_2 there's no more 'spatial similarity' within X than there is under G_1; hence view classes with respect to $(G_2, \mathscr{D}_2)$ will be no more ambiguous about what their realizations could be than those with respect to $(G_1, \mathscr{D}_1)$, and we can expect the ability of $(G_2, \mathscr{D}_2)$ to distinguish objects by view classes to be at least as good as that of $(G_1, \mathscr{D}_1)$. Of course, in the case $G_1 = G_2$, $(G_1, \mathscr{D}_1) \lesssim (G_2, \mathscr{D}_2) \Leftrightarrow \mathscr{D}_1 \lesssim \mathscr{D}_2$. We say $(G_1, \mathscr{D}_1)$ and $(G_2, \mathscr{D}_2)$ are **equivalent**, written $(G_1, \mathscr{D}_1) \sim (G_2, \mathscr{D}_2)$, if $(G_1, \mathscr{D}_1) \lesssim (G_2, \mathscr{D}_2)$ and $(G_2, \mathscr{D}_2) \lesssim (G_1, \mathscr{D}_1)$, equivalently if $G_1 = G_2$ and $\mathscr{D}_1 \sim \mathscr{D}_2$; in other words we get an equivalent similarity view structure just by replacing the *view* structure with an equivalent one. The relation $\lesssim$ is a preordering and $\sim$ is the induced equivalence relation.

2.5 Proposition Let $(G_1, \mathscr{D}_1) \precsim (G_2, \mathscr{D}_2)$ be similarity view structures on X, and Ob be an object set closed under G_1 (hence also G_2). Then for $A,B \in Ob$,

(1) $A \to B$ w.r.t. $(G_2, \mathscr{D}_2) \Rightarrow A \to B$ w.r.t. $(G_1, \mathscr{D}_1)$

(2) $A \leftrightarrow B$ w.r.t. $(G_2, \mathscr{D}_2) \Rightarrow A \leftrightarrow B$ w.r.t. $(G_1, \mathscr{D}_1)$

Proof:

(1) Then for any $D_1 \in \mathscr{D}_1$, where $D_2 \in \mathscr{D}_2$ covers D_1 and $g \in G_2$ embeds D_2A in B, $g \in G_1$ also embeds D_1A in B. Thus, with respect to $(G_1, \mathscr{D}_1)$, every view of A is embedded in B, i.e. $A \to B$.

(2) By (1). o

Putting (2) in the contrapositive, if A and B are view class distinguishable with respect to $(G_1, \mathscr{D}_1)$, they're also view class distinguishable with respect to $(G_2, \mathscr{D}_2)$, thus illustrating the remark made earlier about the relative distinguishing powers of $\precsim$-comparable similarity view structures. It also follows from the above result that the relations $\to$ and $\leftrightarrow$ are equivalence invariant, i.e. if $(G_1, \mathscr{D}_1)$ and $(G_2, \mathscr{D}_2)$ are equivalent then

(1) $A \to B$ w.r.t. $(G_1, \mathscr{D}_1) \Leftrightarrow A \to B$ w.r.t. $(G_2, \mathscr{D}_2)$

(2) $A \leftrightarrow B$ w.r.t. $(G_1, \mathscr{D}_1) \Leftrightarrow A \leftrightarrow B$ w.r.t. $(G_2, \mathscr{D}_2)$.

3. View Self-Similarity

This section is concerned with the visual notion of self-similarity for the expression of which the formal idea of a 'view' was originally developed. After the relevant definitions and a few remarks we'll concentrate on properties of 'view self-similar' closed subsets of a topological space.

The definitions which follow are made with respect to a similarity view space $(X, G, \mathscr{D})$ and object set Ob, for *nonempty* objects A and B. At least in the case where $\mathscr{D}$ is covering and ideal, the definitions $u \rightrightarrows B$, $A \rightrightarrows B$ and $A \leftrightarrows B$ below are powerful strengthenings of the earlier definitions $u \to B$, $A \to B$ and $A \leftrightarrow B$, as the terminology will reflect.

We say a view u is <u>universally embedded</u> in B, written $u \rightrightarrows B$, if u is embedded in every nonempty view of B. Intuitively this means that wherever you look at B ('at' implying a *nonempty* view) you can see a copy of u within your view. As long as $\mathscr{D}$ is covering (so there *exists* a nonempty view of B), this trivially implies that $u \to B$.

We now say A is <u>universally view-embedded</u> in B, written $A \rightrightarrows B$, if every nonempty view of A is universally embedded in B. If $\mathscr{D}$ is covering and ideal, every view of A can be expanded to a *nonempty* view of A, hence the above condition $A \rightrightarrows B$ moreover implies that *every* view of A is universally embedded in B, which in turn implies that $A \to B$. As long as $\mathscr{D}$ is covering, the relation $\rightrightarrows$ is transitive. However, it certainly need *not* be reflexive. If $A \rightrightarrows A$ we say A is <u>view self-similar</u> (abbreviated hereafter to 'self-similar'). This then is a formal expression of the idea that 'wherever you look at A you can see the same structure', saying that every two nonempty views of A are embedded in one another. A simple example is that of a line A in $\mathbb{R}^2$, the latter having the usual similarity view structure.

A and B are <u>universally view-similar</u>, written $A \leftrightarrows B$, if $A \rightrightarrows B$ and $B \rightrightarrows A$. Note that by transitivity of $\rightrightarrows$ this implies A and B are both self-similar. As long as $\mathscr{D}$ is covering and ideal it also implies $A \leftrightarrow B$.

Two further definitions are of relevance. We say A is <u>view-embedded</u> in v, written $A \to v$, if every view of A is embedded in v. If also v is a view of A we'll say v is a <u>generative view of A</u>, since $(\mathscr{D}A)^{\sim} = \{u^{\sim} \mid u \le v\}$, expressing that the view classes of A are 'generated' from the subviews of v. Such a view precisely embodies the visual structure of A modulo similarity. The relevance to self-similarity is that . .

<u>3.1</u> <u>Note</u> If $(G, \mathscr{D})$ is a similarity view structure on X with $\mathscr{D}$ covering and ideal, then for a nonempty object A,

A is self-similar $\Leftrightarrow$ Every nonempty view of A is a generative view of A.

Proof:

Since $\mathscr{D}$ is covering and ideal we know that A is self-similar iff every view of A is embedded in every nonempty view of A, i.e. every nonempty view of A is a *generative* view of A. ○

3.2 Proposition Let $(G_1, \mathscr{D}_1) \lesssim (G_2, \mathscr{D}_2)$ be similarity view structures on X, and Ob be an object set closed under G_1 (hence also G_2). Then for $A,B \in Ob$,

(1) $A \rightrightarrows B$ w.r.t. $(G_2, \mathscr{D}_2) \Rightarrow A \rightrightarrows B$ w.r.t. $(G_1, \mathscr{D}_1)$.

(2) $A \leftrightarrows B$ w.r.t. $(G_2, \mathscr{D}_2) \Rightarrow A \leftrightarrows B$ w.r.t. $(G_1, \mathscr{D}_1)$.

Proof:
(1) Let D_1A and E_1B be nonempty $\mathscr{D}_1$-views of A and B ; we must show some element of G_1 embeds D_1A in E_1B . Since $\mathscr{D}_1 \lesssim \mathscr{D}_2$ let $D_2 \in \mathscr{D}_2$ with $D_1 \subseteq D_2$, and let $E_2 \in \mathscr{D}_2$ with $E_2 \subseteq E_1$ and E_2B nonempty. As $A \rightrightarrows B$ with respect to $(G_2, \mathscr{D}_2)$ let $g \in G_2$ embed D_2A in E_2B . Then $g \in G_1$ also embeds D_1A in E_1B .
(2) By (1). ○

3.3 Corollary The relations $\rightrightarrows$ and $\leftrightarrows$ are equivalence invariant, as is the property of being self-similar. ○

Regarding the possible self-similarity of X itself,

3.4 Note X is self-similar $\Leftrightarrow$ Every view domain is embedded in every other.
Proof:
Since every view of X is nonempty, X is self-similar iff every view of X is embedded in every other. And for $D,E \in \mathscr{D}$, $DX \rightarrow EX \Leftrightarrow D \rightarrow E$. ○

For the rest of this section let $(X,\mathcal{O})$ be a Hausdorff topological space on which we have a similarity view structure $(G, \mathscr{D})$ with $\mathscr{D}$ ideal, and take the object set to be $\mathscr{C}X$, the set of closed subsets of X. We'll now consider some properties of self-similar closed sets. The first result gives a prime characteristic of such sets, namely that they're residual, i.e. they have empty interior. Bearing in mind that $\mathscr{D}$ is covering and ideal, recall that an (implicitly nonempty) object is self-similar iff every view of A is embedded in every nonempty view of A.

3.5 Proposition Every self-similar proper closed subset A of X is residual.
Proof:
As $A \neq X$ there's a non-full view u of A, and since u is embedded in every nonempty view of A, no view of A can be full, i.e. no view domain is a subset of A. As the view domains form a basis for the topology, this just says that A has empty interior. ○

X is thus the only closed set which could possibly be self-similar with nonempty interior. A second characteristic of most self-similar closed sets is that of being 'perfect', i.e. each point of the set is a cluster point of the set ;

3.6 **Proposition** Every self-similar closed set A with at least two points is perfect.

Proof:

As $\mathscr{D}$ is covering and ideal and $|A| \geq 2$, there's a view u of A whose domain contains at least two points of A. Now take any $x \in A$. Then for all $D \in \mathscr{D}$ with $x \in D$, as $u \to DA$ (and similarities are injective) then D contains at least two points of A, hence at least one point of $A - \{x\}$; so as $\mathscr{D}$ is a basis for the topology, x is a cluster point of A as required. □

The next result concerns self-similar *compact* sets, showing that they're generally somewhat disconnected. You might see why this could be expected by considering the usual similarity view structure on $\mathbb{R}^n$. There, any self-similar compact set A is bounded and can thus be covered by a view domain, which gives rise to a view of A having a 'moat' of background space surrounding A ; and since this view is embedded in every nonempty view of A we can expect A to contain 'islands surrounded by moats' everywhere you look, and A should thus be very disconnected (as long as A is non-trivial). This 'island embedding' idea forms the basis of the proof ;

3.7 **Proposition** If X is regular, every self-similar compact set A with $|A| \geq 2$ has at least continuum-many components (moreover, continuum-many components in any open set intersecting A).

Proof:

As A is compact with $\mathscr{D}$ ideal, A is covered by some $D \in \mathscr{D}$, and as $|A| \geq 2$ and X is Hausdorff there are disjoint view domains E_1 and E_2 intersecting A with $E_i \subseteq D$. By self-similarity of A let g_1 and g_2 be similarities respectively embedding DA in E_1A and E_2A. In particular they embed DA in DA, and $g_i A \subseteq A$.

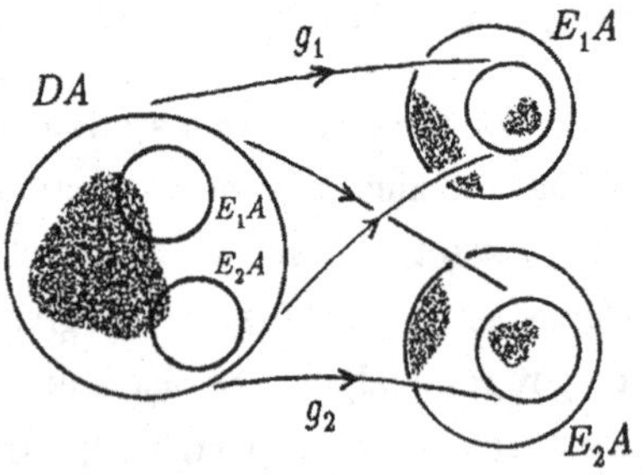

For any finite sequence g in $\{g_1, g_2\}$ we have the composition $\circ g$ which likewise embeds DA in DA (by transitivity of embedding) so $g A = A \cap g D$, and since $g D$ is an open neighbourhood of the compact set $g A$ and X is regular, $g A$ is a union of components of A.

Now for $g \in \{g_1, g_2\}^\omega$, $\langle g \restriction n \, A \rangle$ is a decreasing sequence of nonempty compact sets hence has nonempty compact intersection which we'll denote by $g A$. For *distinct* $f, g \in \{g_1, g_2\}^\omega$, where n is least such that $f_n \neq g_n$ (so $f \restriction n = g \restriction n$) we have

$$f A \subseteq f \restriction n{+}1 \, A \subseteq f \restriction n{+}1 \, D = f \restriction n (f_n D) \subseteq f \restriction n E_1 \quad \text{and}$$
$$g A \subseteq g \restriction n{+}1 \, A \subseteq g \restriction n{+}1 \, D = f \restriction n (g_n D) \subseteq f \restriction n E_2$$

with $f \restriction n E_1$ and $f \restriction n E_2$ disjoint as E_1 and E_2 are, so $f \restriction n{+}1 \, A$ and $g \restriction n{+}1 \, A$ are disjoint; and remember they're each unions of components of A. It follows that any component of A intersecting $f A$ (and there *is* one of course) is disjoint from any component of A intersecting gA. Hence A has at least as many components as elements of $\{g_1, g_2\}^{\omega}$, i.e. at least $2^{\aleph_0}$ components.

Moreover, consider any open set V intersecting A. There's $E \in \mathcal{D}$ intersecting A with $E \subsetneq V$, and since DA (above) is embedded in EA by some similarity g, the components of A map via g to components of A in E, hence in V. So at least $2^{\aleph_0}$ components of A lie in V. ○

The *existence* of non-trivial self-similar compact sets in the case of $\mathbb{R}^n$ with the usual similarity view structure, perhaps surprising at first (you may like to try and think of an example . . .), is demonstrated in the next section after 4.7. Trivially every *singleton* is self-similar in the case of $\mathbb{R}^n$. However, as the last result shows, to find *connected* non-trivial self-similar closed sets we must look in the realm of *unbounded* closed sets. In the next section we'll provide a class of examples obtained by extending certain of Hutchinson's invariant sets.

4. Self-Similarity of Some ω-Extensions of Invariant Sets

Throughout let X be a nonempty locally compact (hence complete) metric space. Since X is locally compact, the body-interiors (see 'Preliminaries') form a *basis* for the topology of X (using that each point of X has a neighbourhood basis of compact sets, and the interior of any compact set is a subcompact regular open set). So let $\mathscr{D}$ be the view structure on the space X consisting of the body-interiors, and note that $\mathscr{D}$ is ideal. Using 2.4, note that $\mathscr{D}$ is equivalent to the (maximal) view structure consisting of the nonempty subcompact open sets (which, if X is boundedly compact, is *also* equivalent to the view structure consisting of the open balls). In the terminology of Chapter 3, let F be an admissible set of contractions of X with invariant set K, and let $hF = \{f \in F \mid f \text{ is a homeomorphism of } X\}$ and G_F be the group of homeomorphisms of X generated by hF (this is the trivial group if $hF = \emptyset$). Then with respect to the similarity view structure $(G_F, \mathscr{D})$ on X we'll give sufficient conditions for the self-similarity of certain 'ω-extensions' of K. Such ω-extensions will then also be self-similar with respect to any *weaker* similarity view structure of course; in particular, if $X = \mathbb{R}^n$ and every element of hF is a direct similitude, the sets in question will be self-similar with respect to the usual similarity view structure on $\mathbb{R}^n$.

First we'll describe what 'ω-extensions' of K are, and why in suitable cases they look like being good prospects for self-similarity. Consider the example of a Sierpinski Gasket K in $\mathbb{R}^2$ produced by $F = \{f_1, f_2, f_3\}$ as described in Section 1 of Chapter 3. By visual inspection one can soon convince oneself that any two views of K lying fully *within* the triangle involved can be embedded in one another, moreover that the embeddings can be done by elements of the monoid generated by F. Roughly speaking then, the 'inside' of K is self-similar, and it is only due to the presence of an outer edge of K that K itself is not self-similar (for example, no *whole* view of K is embedded in the inside of K; indeed we *know* by 3.7 that K is not self-similar, since K is compact and connected). If we could in some way 'grow' or 'extend' K outwards indefinitely so that there was no edge left then, we might well end up with a genuine self-similar set. This idea of extending K outwards, hinted at in [Man], is the basic idea behind forming an 'ω-extension', which, returning to the general case, we now formally describe.

For $f \in hF$, as K is closed under f it's expanded by f^{-1}, i.e. $K \subseteq f^{-1}K$. Consequently, for any infinite sequence $\boldsymbol{i}$ of inverses of elements of hF, $\langle \boldsymbol{i}{\restriction}n\, K\rangle$ is an *increasing* sequence of compact sets (the notation here being the same as that used in Chapter 3, namely $\boldsymbol{i}{\restriction}n\, K = \circ(\boldsymbol{i}{\restriction}n)\, K = (\boldsymbol{i}_0 \circ \cdots \circ \boldsymbol{i}_{n-1})\, K$, with $\boldsymbol{i}{\restriction}0\, K = K$) and we define $\boldsymbol{i}K = \bigcup_n \boldsymbol{i}{\restriction}n\, K$. A set of this form is called an <u>ω-extension of K with respect to F</u>, abbreviated to 'ω-extension of K' with F in mind. We similarly define $\boldsymbol{i}A = \bigcup_n \boldsymbol{i}{\restriction}n\, A$ for any $A \subseteq X$ closed under all $f \in hF$ (so $\langle \boldsymbol{i}{\restriction}n\, A\rangle$ is increasing).

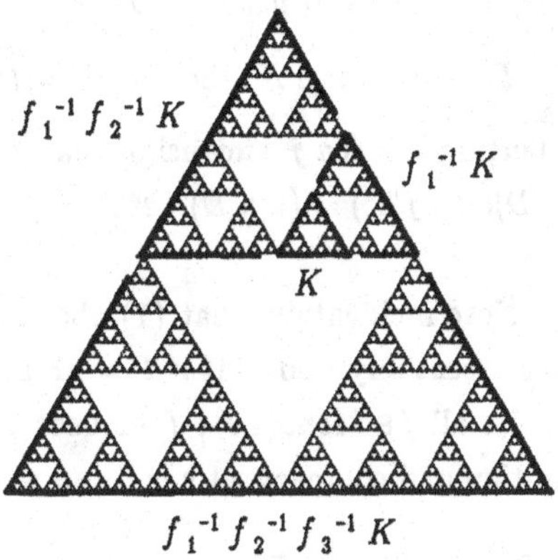

The sketch on the right illustrates the first few stages in forming an ω-extension of a Sierpinski Gasket, which should be of help in visualizing the way in which the extending takes place. Note for example that $f_1^{-1} f_2^{-1} f_3^{-1} K$ is to K as K is to $f_3 f_2 f_1 K$.

Letting $(hF)^{-1}$ denote the set of inverses of elements of hF, we'll denote the set of infinite sequences in $(hF)^{-1}$ by $(hF)^{-\omega}$, the set of finite sequences in $(hF)^{-1}$ by $(hF)^{-<\omega}$, and the set of sequences of length n in $(hF)^{-1}$ by $(hF)^{-n}$. A set of the form $\boldsymbol{i} A$ where $\boldsymbol{i} \in (hF)^{-n}$ may be called a <u>$(-n)$th-level image of A</u> with respect to hF; it is obtained from A by applying a sequence of n elements taken from $(hF)^{-1}$.

To prove self-similarity of an ω-extension $\boldsymbol{i} K$ we'll require two conditions on F, namely the homeomorphism condition and the compact set condition introduced respectively in Sections 6 and 7 of Chapter 3, and one condition jointly dependent on $\boldsymbol{i}$ and F which relates $\boldsymbol{i}$ to the way in which the compact set condition can hold, making sure that the ω-extension in question extends over the *whole* space in a certain sense, thus ensuring that it has no 'outside'.

In the following, D will always be assumed to be an element of $\mathscr{D}$. Bear in mind that if every element of F is a homeomorphism then the recurring phrase 'If the compact set condition holds with $\overline{D}$' in the work below can be replaced by 'If the open set condition holds with D', by 7.5 of Chapter 3.

<u>**4.1**</u> <u>**Proposition**</u> If F satisfies the compact set condition with $\overline{D}$ then

(1) Every $f \in$ *Monoid* hF embeds DK in DK.

(2) $\forall \boldsymbol{i} \in (hF)^{-<\omega}$, $D\,\boldsymbol{i}\,K = DK$.

(3) $\forall \boldsymbol{i} \in (hF)^{-\omega}$, $D\,\boldsymbol{i}\,K = DK$.

(4) $\forall \boldsymbol{i} \in (hF)^{-\omega}$ $\forall n$, $\circ(\boldsymbol{i}{\restriction}n)$ embeds DK in $\boldsymbol{i}\,K$.

Proof:

(1) Let $H = \overline{D}$. By transitivity of embedding it suffices to show that each $f \in hF$ embeds DK in DK. As H is closed under the homeomorphism f so is its interior D. It remains to show that $f(K \cap D) = K \cap f D$ so that $f(DK) = (f D)K$. For $g \in F - \{f\}$, $g\,H$ is disjoint from $\underline{f H} = f \underline{H} = f D$ so (as $K \subseteq H$ gives $g\,K \subseteq g\,H$) $g\,K$ is disjoint from $f D$ too. So as K is the union of its first-level images, $K \cap f D = f K \cap f D = f(K \cap D)$ as required.

(2) Let $g = \circ\boldsymbol{i}$. Then $g^{-1} \in$ *monoid* hF so by (1) $g^{-1}(DK) = (g^{-1}D)K$, so applying g to both sides, $DK = g((g^{-1}D)K) = (g\,g^{-1}D)\,g\,K = D\,g\,K = D\,\boldsymbol{i}\,K$.

(3) By (2), $\forall n \;\; D \cap \boldsymbol{i}{\restriction}n\, K = D \cap K$. Hence $D \cap \boldsymbol{i}K = D \cap \bigcup_n \boldsymbol{i}{\restriction}n\, K = \bigcup_n (D \cap \boldsymbol{i}{\restriction}n\, K)$ $= \bigcup_n (D \cap K) = D \cap K$. So $D\,\boldsymbol{i}\,K = DK$.

(4) Letting $\boldsymbol{i} = \boldsymbol{i}{\restriction}n\; \boldsymbol{j}$ and using that $D\,\boldsymbol{j}\,K = DK$ by (3), $\boldsymbol{i}{\restriction}n\,(DK) = \boldsymbol{i}{\restriction}n\,(D\,\boldsymbol{j}\,K) =$ $(\boldsymbol{i}{\restriction}n\; D)(\boldsymbol{i}{\restriction}n\; \boldsymbol{j}K) = (\boldsymbol{i}{\restriction}n\; D)\; \boldsymbol{i}\,K$. ○

Note incidentally that (1) above gives an alternative proof of 7.10 of Chapter 3, which essentially said that if $F = hF$ and F satisfies the compact set condition with $\overline{D}$, then for all $f \in$ *Monoid* F, f embeds DK in K. Now combining the homeomorphism condition and compact set condition we have the following ;

4.2 **Proposition** If F satisfies the homeomorphism condition and satisfies the compact set condition with $\overline{D}$, then

(1) DK is embedded in every nonempty view of K.

(2) For any nonempty subviews u and v of DK, $u \rightarrow v$.

Proof:

(1) Let EK be a nonempty view of K. Taking $x \in E \cap K$, by 6.1(2) of Chapter 3 there's $f \in$ *Semigroup* hF with $fD \subseteq E$; and we know f embeds DK in DK by 4.1(1), so f also embeds it in EK, $f(DK) = (fD)K$ being a subview of EK.

(2) By (1) DK is embedded in v; hence so is the subview u of DK. ○

4.3 **Lemma** For $D \in \mathscr{D}$ closed under all $f \in hF$, and $\boldsymbol{i} \in (hF)^{-\omega}$,

$$\boldsymbol{i}\,D = X \Leftrightarrow \forall E \in \mathscr{D} \;\; \exists n \in \omega \;\;\; E \subseteq \boldsymbol{i}{\restriction}n\, D.$$

Proof:

$\Rightarrow$: For $E \in \mathscr{D}$, $\overline{E}$ is compact and covered by the open sets $\boldsymbol{i}{\restriction}n\, D$ since their union is $\boldsymbol{i}\,D = X$. Hence there's a finite subcover, and this posesses a *largest* member since $\langle \boldsymbol{i}{\restriction}n\; D\rangle$ is increasing.

$\Leftarrow$: Trivial, as $\mathscr{D}$ covers X and $\forall n \in \omega \;\; \boldsymbol{i}{\restriction}n\, D \subseteq \boldsymbol{i}\,D$. ○

We now have the main result ;

4.4 **Proposition** (Self-Similarity of Some ω-Extensions $\boldsymbol{i}\,K$).

If F satisfies the homeomorphism condition and satisfies the compact set condition with $\overline{D}$, then for any $\boldsymbol{i} \in (hF)^{-\omega}$ such that $\boldsymbol{i}\,D = X$, $\boldsymbol{i}\,K$ is a self-similar closed set (and DK is a generative view of $\boldsymbol{i}\,K$).

Proof:

First we show $\boldsymbol{i}\,K$ is closed. Take any $x \in \overline{\boldsymbol{i}\,K}$, and let $n \in \omega$ with $x \in \boldsymbol{i}{\restriction}n\, D$. By 4.1(4), $\boldsymbol{i}\,K \cap \boldsymbol{i}{\restriction}n\, D = \boldsymbol{i}{\restriction}n\,(K \cap D) = \boldsymbol{i}{\restriction}n\, K \cap \boldsymbol{i}{\restriction}n\, D$, so as $\boldsymbol{i}{\restriction}n\, D$ is open, $x \in \overline{\boldsymbol{i}{\restriction}n\, K} =$ $\boldsymbol{i}{\restriction}n\, K \subseteq \boldsymbol{i}\,K$, giving $x \in \boldsymbol{i}\,K$ as required.

Every view of $\boldsymbol{i}\,K$ is embedded in DK (*Proof*: for $E \in \mathscr{D}$, by 4.3 there's $n \in \omega$ with $E \subseteq \boldsymbol{i}{\restriction}n\, D$, hence $E\,\boldsymbol{i}\,K$ is a subview of $(\boldsymbol{i}{\restriction}n\; D)\boldsymbol{i}\,K$ which by 4.1(4) is $\boldsymbol{i}{\restriction}n\,(DK)$ which is similar to DK; so $E\,\boldsymbol{i}\,K$ is embedded in DK) so every two nonempty views u and v

of $\boldsymbol{i}\,K$ are similar to nonempty subviews u' and v' of DK, and by 4.2(2) we know that $u' \to v'$, hence $u \to v$. So $\boldsymbol{i}\,K$ is self-similar. And lastly, since DK is a view of $\boldsymbol{i}\,K$ (see 4.1(3)) the above also showed that DK is a *generative* view of $\boldsymbol{i}\,K$. ○

For example, let $X = \mathbb{R}^2$ and K be a Sierpinski Gasket produced by the usual F. Let D be the interior of the triangle involved and let $\boldsymbol{i} \in F^{-\omega}$ such that every element of F^{-1} occurs infinitely often in $\boldsymbol{i}$. It is easily seen that $\boldsymbol{i}\,D = X$ in such a case (indeed, these are the *only* $\boldsymbol{i}$ giving $\boldsymbol{i}\,D = X$), so by 4.4 $\boldsymbol{i}\,K$ is a self-similar closed set. In addition, by the following result, every two such ω-extensions are universally view-similar (hence also view-class indistinguishable) ;

4.5 Corollary (Universal View-Similarity Among ω-Extensions of K).

If F satisfies the homeomorphism condition and satisfies the compact set condition with $\overline{D}$, then for any $\boldsymbol{i}, \boldsymbol{j} \in (hF)^{-\omega}$ such that $\boldsymbol{i}\,D = X = \boldsymbol{j}\,D$, $\boldsymbol{i}\,K \leftrightarrows \boldsymbol{j}\,K$.

Proof:

By the last result DK is a generative view of both $\boldsymbol{i}\,K$ and $\boldsymbol{j}\,K$. So for any nonempty view u of $\boldsymbol{i}\,K$ and nonempty view v of $\boldsymbol{j}\,K$, u and v are similar to nonempty subviews u' and v' of DK, and by 4.2(2) $u' \to v'$, hence $u \to v$. ○

Regarding 4.4, the following result describes a class of *periodic* sequences $\boldsymbol{i}$ having the desired property that $\boldsymbol{i}\,D = X$ in the case where $X = \mathbb{R}^n$ and the elements of F are similitudes. We define first that for $f \in (hF)^n$, $f^{-1} = \langle f_{n-1}^{-1}, \ldots, f_0^{-1} \rangle \in (hF)^{-n}$, so $\circ(f^{-1}) = (\circ f)^{-1}$, and for $m \leq \omega$ we define $f^{-m} = (f^{-1})^m$.

4.6 Note Suppose $X = \mathbb{R}^n$, every element of F is a similitude, and F satisfies the compact set condition with $A = \overline{D}$. Then for any $f \in F^{<\omega}$ such that $fA \subsetneq \underline{A}$, letting $\boldsymbol{i} = f^{-\omega}$ we have $\boldsymbol{i}\,D = X$, so $\boldsymbol{i}\,K$ is a self-similar closed set.

Proof:

Since $fA \subsetneq D$ with fA compact, there's $\epsilon > 0$ with $[fA]_\epsilon \subsetneq D$. Letting λ be the scale factor of $\circ(f^{-1})$, then for all $m \geq 1$ we have (noting $A \subsetneq f^{-(m-1)}A = f^{-m}fA$) $[A]_{(\lambda^m\epsilon)} \subsetneq [f^{-m}fA]_{(\lambda^m\epsilon)} = f^{-m}[fA]_\epsilon \subsetneq f^{-m}D \subsetneq \boldsymbol{i}\,D$. Since $\lambda^m\epsilon$ converges to ∞ with m, it follows that $\boldsymbol{i}\,D = X$, and by 4.4 $\boldsymbol{i}\,K$ is a self-similar closed set. ○

Finally we give sufficient conditions for K *itself* to be self-similar. As well as the homeomorphism condition and compact set condition for some $\overline{D}$, we stipulate a condition ensuring that arbitrarily large views of K are embedded in DK, which by 4.2(1) will imply the self-similarity of K. Regarding the term 'arbitrarily large' used in (2) below, when we say 'arbitrarily large' elements of $\mathscr{D}$ obey a certain condition, we mean that for all $D \in \mathscr{D}$ there's $E \in \mathscr{D}$ with $D \subsetneq E$ such that E obeys the condition.

4.7 Proposition (Self-Similarity of Some K).

If F satisfies the homeomorphism condition and satisfies the compact set condition with $\overline{D}$ such that, for arbitrarily large $E \in \mathscr{D}$ there's $f \in hF$ with $fE \subseteq D$ and fE disjoint from gD for all $g \in F - \{f\}$, then K is self-similar (and DK is a generative view of K).

Proof:

By 4.2(1) and transitivity of embedding, it suffices to show that every view of K is embedded in DK (i.e. that DK is a generative view of K). Taking any $C \in \mathscr{D}$ then, we wish to show that CK is embedded in DK. For some $E \in \mathscr{D}$ with $C \subseteq E$, there's $f \in hF$ with $fE \subseteq D$ and fE disjoint from gD for all $g \in F - \{f\}$. Taking such E, it now suffices to show that EK is embedded in DK, for then CK will be too. As $fE \subseteq D$, it remains to prove $f(K \cap E) = K \cap fE$, i.e. $fK \cap fE = K \cap fE$. And this holds as K is the union of its first-level images and gD is disjoint from fE for all $g \in F - \{f\}$. $\circ$

For example, consider the case in $\mathbb{R}$ with $F = \{f_n \mid n \leq \omega\}$, where for $n < \omega$ f_n is the direct similitude mapping $A = [0,1]$ to $[a_n, b_n] = [1/2^n - 1/4^{n+1}, 1/2^n] \subseteq A$ (i.e. f_n contracts about 1 by factor $r_n = 1/4^{n+1}$ then translates by $-1 + 1/2^n$), whilst f_ω is constant value 0. Note that $f_\omega = \lim \langle f_n \rangle$, so F is compact and satisfies the homeomorphism condition. The first-level images of A are shown in the diagram. Letting $c_n = 2^{n+1} - 1$, a simple calculation shows that for $n \geq 1$, the closed $c_n r_n$ fringe of $f_n A$ (namely $[a_n - c_n r_n, b_n + c_n r_n]$, which is $f([A]_{c_n})$) is nonoverlapping with gA for all $g \in F - \{f\}$. Since $\langle c_n \rangle$ tends to infinity, it's easily seen that the conditions stipulated in 4.7 hold with $D = \underline{A} = (0,1)$ as we can take the 'arbitrarily large $E \in \mathscr{D}$' to be the interiors of the sets $[A]_{c_n}$. So K is self-similar with respect to the similarity group generated by hF, and is thus also self-similar with respect to the usual similarity view structure on $\mathbb{R}$. We thus have an example of a *bounded* closed set which is self-similar with respect to the usual view structure on $\mathbb{R}$ (and examples in $\mathbb{R}^2$ can now easily be produced using a similar scheme to the one underlying the above). Establishing the existence of such a set was the main reason we introduced the homeomorphism condition (in Chapter 3) rather than keeping to the simpler condition that every element of F be a homeomorphism. For, in the case where $X = \mathbb{R}^n$, if F satisfies the conditions stipulated in 4.7 then F cannot be a set of similitudes. Indeed, if every element of hF is a similitude then F must contain a constant map. We can prove this nonstandardly as follows. By transfer of the condition on F (and bearing in mind that as $\mathscr{D}$ is an ideal basis, *some* element of $\mathscr{D}^*$ expands $\nu\mathscr{D} = \bigcup\{D^* \mid D \in \mathscr{D}\} = bd\, X^*$) for some $E \in \mathscr{D}^*$ expanding $bd\, X^*$ there's $f \in hF^*$ with $fE \subseteq D$, which implies that the scale factor of f is infinitesimal. It follows that the element of F to which f is near must be a constant map.

5. The View Topology

Throughout let X be a locally compact Hausdorff space and $\mathscr{D}$ be a view structure on the space X consisting of body-interiors. $\mathscr{D}$ *could* consist of *all* such sets (since they indeed form a basis for X as X is locally compact), but this is not required. The main example we have in mind is where X is $\mathbb{R}^n$ and $\mathscr{D}$ is the usual view structure on $\mathbb{R}^n$, consisting of the open balls. Our object set will be $\mathscr{C}X$, the set of closed subsets of X. The set of views then is $\mathscr{V} = \mathscr{D}\,\mathscr{C}X = \{DA \mid D \in \mathscr{D} \text{ and } A \in \mathscr{C}X\} = \{(D,I) \mid D \in \mathscr{D} \text{ and } I \text{ is a closed-in-}D \text{ subset of } D\}$ (bearing in mind that each I in the latter is an intersection of D with some $A \in \mathscr{C}X$).

In this section we describe and investigate three topologies. First of all we'll consider the 'view topology' on $\mathscr{V}$, which will then give rise to the 'view-induced' topology on $\mathscr{C}X$. Then we'll add in the idea of similarities so we moreover have a *similarity* view structure on X , thus bringing the concept of view classes into the fray, and we'll show how the view topology naturally induces a topology on the set $\mathscr{V}^{\sim}$ of view classes.

We start then by considering how to define a natural topology on $\mathscr{V}$. The rough idea is that, where $(D,I) \in \mathscr{V}$ and $(E,J) \in \mathscr{V}^*$, (E,J) should be near (D,I) iff E is near D and J looks like I^*. The first of these requirements suggests we need a suitable topology on the set $\mathscr{D}$ of view domains. To this end then, consider the following nonstandard anatomy of a view domain D. As $\overline{D}$ is compact with interior D, $\partial D = \overline{D} - D = \partial\,\overline{D}$, which is compact as $\overline{D}$ is. Note then that $\overline{D}$ is the disjoint union of D and ∂D, and therefore $\overline{D}^{\,\mu}$ is the disjoint union of D^{μ} and $(\partial D)^{\mu}$, which are each partitioned into monads of course. This is illustrated on the right. Note also that $D^{\mu} \subseteq D^* \subseteq \overline{D}^{\,\mu}$, i.e. D^* is sandwiched between D^{μ} and $D^{\mu} \cup (\partial D)^{\mu}$. Regarding the question of when $E \in \mathscr{D}^*$ should be near D, it seems natural to ask that E be sandwiched between D^{μ} and $\overline{D}^{\,\mu}$ like D^*, i.e. that $D^{\mu} \subseteq E \subseteq \overline{D}^{\,\mu}$, so that E differs from D^* only within the 'infinitesimally thin' boundary region $(\partial D)^{\mu}$. Such a topology on $\mathscr{D}$ indeed exists; namely the body-interior topology described in Appendix 6. We shall assume from now on then that $\mathscr{D}$ has this topology, so that for $D \in \mathscr{D}$ and $E \in \mathscr{D}^*$,

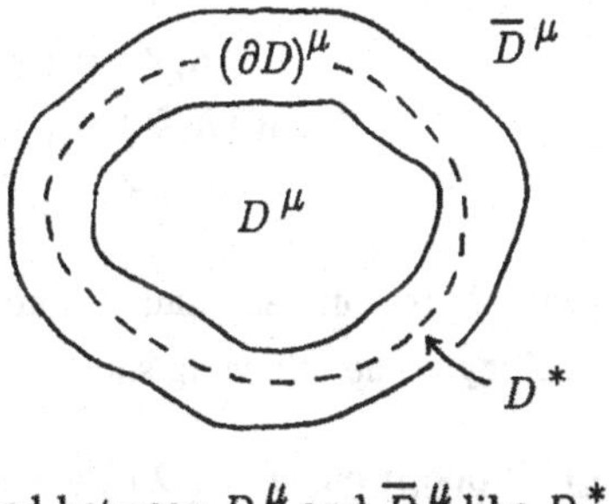

$$E \in \mu D \Leftrightarrow D^{\mu} \subseteq E \subseteq \overline{D}^{\,\mu} \Leftrightarrow E \,\Delta\, D^* \subseteq (\partial D)^{\mu}.$$

The second part of the problem, returning to $(D,I) \in \mathscr{V}$ and $(E;J) \in \mathscr{V}^*$ and assuming now that $E \in \mu D$, is to specify when J 'looks like' I^*. Intuitively, the only part of D^* we can *properly* see (assuming we cannot resolve beyond the monadic level, i.e. that points in the same monad cannot be distinguished) is the main body D^{μ}; the *remaining* points belong to $(\partial D)^{\mu} = \bigcup\{\mu x \mid x \in X$ and μx contains a point of $\partial(D^*) = (\partial D)^*\}$ and are thus too near the boundary of D^* to be resolved from it. So our task reduces to specifying when J 'looks like' I^* within D^{μ}. Since we cannot resolve monads, $J \cap D^{\mu}$ blurs to $\bigcup\{\mu x \mid x \in D$ and μx contains an element of $J\}$, shown below, and likewise $I^* \cap D^{\mu}$ blurs to $\bigcup\{\mu x \mid x \in D$ and μx contains an element of $I^*\}$;

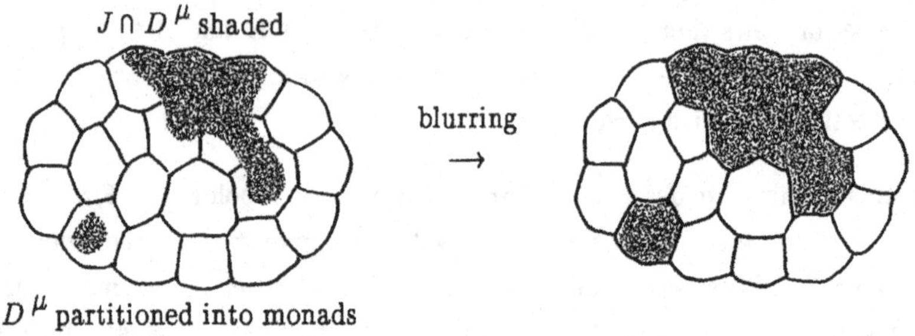

Intuitively then, J looks like I^* in D^{μ} iff these blurred images agree, i.e. $\forall x \in D$, J intersects μx iff I^* intersects μx. Since J intersects μx iff $x \in st\, J$, and I^* intersects μx iff $x \in I$, the condition is equivalently that $\forall x \in D$, $x \in st\, J \Leftrightarrow x \in I$; equivalently $(st\, J) \cap D = I$. So, we now ask whether there *is* a topology on $\mathscr{V}$ in which $(E,J) \in \mu(D,I) \Leftrightarrow E \in \mu D$ and $(st\, J) \cap D = I$. The answer is yes, and we now give a standard description.

The <u>**view topology**</u> on $\mathscr{V}$ is the topology generated by the sets of the form

$$[K,V] = \{(D,I) \mid K \subseteq D \text{ and } \overline{D} \subseteq V\},$$
$$\mathbf{int}\,(K,U) = \{(D,I) \mid K \subseteq D \text{ and } I \text{ intersects } U \subseteq D\}, \text{ or}$$
$$\mathbf{disj}\,K = \{(D,I) \mid K \subseteq D \text{ and } I \text{ is disjoint from } K\},$$

where K is compact and U and V are open. From now on assume $\mathscr{V}$ has the view topology. The following shows that we have the monads desired ;

<u>**5.1**</u> <u>**Proposition**</u> For $(D,I) \in \mathscr{V}$ and $(E,J) \in \mathscr{V}^*$,

$$(E,J) \in \mu(D,I) \Leftrightarrow E \in \mu D \text{ and } (st\, J) \cap D = I$$
$$\Leftrightarrow E \in \mu D \text{ and } st(J \cap D^{\mu}) = I.$$

Proof:
The two conditions on the right are equivalent since $(st\, J) \cap D = st(J \cap D^{\mu})$. We now show that $(E,J) \in \mu(D,I) \Leftrightarrow E \in \mu D$ and $(st\, J) \cap D = I$.

$\Rightarrow$: By using the sets of the form $[K,V]$ we have that $E \in \mu D$ (recalling a note in Appendix 6 on a basis for the body-interior topology). It remains to show that for $x \in D$, $x \in st\, J \Leftrightarrow x \in I$. First suppose $x \in I$. By local compactness let K be a compact neighbourhood of x with $K \subseteq D$. For any open neighbourhood U of x with $U \subseteq K$, $(D,I) \in \text{int}\,(K,U)$ so $(E,J) \in \text{int}\,(K,U)^*$ so J intersects U^*. By saturation then, J intersects μx, hence $x \in st\, J$ as required. On the other hand suppose $x \notin I$. Then as I is closed in D there's a compact neighbourhood $K \subseteq D$ of x disjoint from I, so $(D,I) \in \text{disj}\, K$ so $(E,J) \in (\text{disj}\, K)^*$ so J is disjoint from K^*, hence from μx, so $x \notin st\, J$ as required.

$\Leftarrow$: (1) If $(D,I) \in [K,V]$ then $(E,J) \in [K,V]^*$ as $E \in \mu D$.

(2) If $(D,I) \in \text{int}\,(K,U)$ then $(E,J) \in \text{int}\,(K,U)^*$ since firstly $K^* \subseteq E$ (as $K^* \subseteq K^\mu \subseteq D^\mu \subseteq E$), and secondly, where $i \in I \cap U$ and (as $i \in st\, J$) $j \in J \cap \mu x$, we have $j \in J \cap U^*$ so J intersects $U^* \subseteq D^*$.

(3) If $(D,I) \in \text{disj}\, K$ then (noting $K^* \subseteq E$ as in (2)) $(E,J) \in (\text{disj}\, K)^*$, otherwise J would *intersect* K^* and hence $I = (st\, J) \cap D$ would intersect $K = st\, K^*$. ○

5.2 Proposition $\mathscr{V}$ is Hausdorff.

Proof:

If $(E,J) \in \mu(D_1, I_1) \cap \mu(D_2, I_2)$, then firstly $E \in \mu D_1 \cap \mu D_2$ so $D_1 = D_2$ as $\mathscr{D}$ is Hausdorff, and then $I_1 = (st\, J) \cap D_1 = (st\, J) \cap D_2 = I_2$; so $(D_1, I_1) = (D_2, I_2)$. ○

For $D \in \mathscr{D}$, the **D-topology on $\mathscr{C}X$** is the topology induced by the map $A \mapsto DA$. Denoting monads with respect to this topology by μ_D then, $B \in \mu_D A \Leftrightarrow D^* B \in \mu\, DA$; or intuitively, B is near A iff B looks like A^* in D^*. The **view-induced topology on $\mathscr{C}X$** is the conjunction of the D-topologies for $D \in \mathscr{D}$, and we'll denote monads with respect to it by μ_s, for reasons shortly to become apparent. Thus, $B \in \mu_s A \Leftrightarrow \forall D \in \mathscr{D}\ B \in \mu_D A$. Equivalently ;

5.3 Proposition For $A \in \mathscr{C}X$ and $B \in \mathscr{C}X^*$,

(1) For $D \in \mathscr{D}$, $B \in \mu_D A \Leftrightarrow (st\, B) \cap D = A \cap D \Leftrightarrow st\,(B \cap D^\mu) = A \cap D$.

(2) $B \in \mu_s A \Leftrightarrow st\, B = A$.

Proof:

(1)
$$\begin{aligned} B \in \mu_D A &\Leftrightarrow D^* B \in \mu\, DA \\ &\Leftrightarrow (D, B \cap D^*) \in \mu(D, A \cap D) \\ &\Leftrightarrow st((B \cap D^*) \cap D^\mu) = A \cap D && \text{by 5.1} \\ &\Leftrightarrow st(B \cap D^\mu) = A \cap D && \text{noting } D^\mu \subseteq D^* \\ &\Leftrightarrow (st\, B) \cap D = A \cap D && \text{as } st\,(B \cap D^\mu) = (st\, B) \cap D. \end{aligned}$$

(2) By (1), since $\mathscr{D}$ covers X. ○

Thus the view-induced topology on $\mathscr{C}X$ turns out to be the 'S-compact' topology described separately in Appendix 7, which is why we've used μ_s for the monads. Furthermore, for $D \in \mathscr{D}$, the 'D-topology' we've defined here is the 'D-topology' defined independently in that appendix. This shows then how the S-compact topology on $\mathscr{C}X$ arises as the natural topology induced by views, and how in particular each D-topology (for $D \in \mathscr{D}$) models 'likeness in D'. Assume $\mathscr{C}X$ has the view-induced topology for the rest of the section.

We could have given *alternative* view-related justification for the desired monads of a view-induced topology on $\mathscr{C}X$ without needing the view topology, as follows. Imagine looking at X^* with vision that is the 'limit' of our standard abilities. Intuitively then, the 'observable part' of X^* is the union monad of $\mathscr{D}$, namely $\bigcup\{D^* \mid D \in \mathscr{D}\} = ns\, X^*$, so for $B \subseteq X^*$ the observable part of B is $B \cap ns\, X^*$. But we cannot resolve beyond the monadic level, i.e. points in the same monad cannot be distinguished, so $B \cap ns\, X^*$ 'blurs' to $\bigcup\{\mu x \mid x \in X$ and μx contains an element of $B\} = \bigcup\{\mu x \mid x \in st\, B\} = (st\, B)^\mu$, which we shall call the <u>monadic image</u> of B, denoted by B^μ (which is $B^{\approx}$ if X is metric and $B \subseteq ns\, X^*$). The natural question arises then of whether there's a topology on $\mathscr{C}X$ in which, for $A \in \mathscr{C}X$ and $B \in \mathscr{C}X^*$, B is near A iff B and A^* have the same monadic image. Now since $st\, A^* = A$, the monadic image of A^* is just the monadic cover of A, hence $B^\mu = A^{*\mu} \Leftrightarrow (st\, B)^\mu = A^\mu \Leftrightarrow st\, B = A$, so the answer to the above question is 'yes', the topology being (again) the S-compact topology on $\mathscr{C}X$. Note that the idea of monadic images was partially used in developing the view topology, since we essentially decided in considering when (E,J) should be near (D,I), that J looks like I^* in D^μ iff the intersection with D^μ of the monadic images of J and I^* were the same (since for any $L \subseteq X^*$, $L^\mu \cap D^\mu = \bigcup\{\mu x \mid x \in D$ and μx contains an element of $L\}$). And note that in the path we took to the view-induced topology on $\mathscr{C}X$, the idea that the 'observable part' of X^* is $\bigcup\{D^* \mid D \in \mathscr{D}\}$ is embedded in our forming the topology as the *conjunction* of the D-topologies for $D \in \mathscr{D}$. Developmental paths aside, these remarks give further perspectives on the topologies defined so far, and from Appendix 7 we know that under the view-induced topology, $\mathscr{C}X$ is a compact Hausdorff space.

<u>5.4</u> <u>Note</u> A $*$-view is nearstandard iff its domain is. In other words, for $(E,J) \in \mathscr{V}^*$, (E,J) is nearstandard $\Leftrightarrow$ E is nearstandard.

Proof:

$\Rightarrow$: Trivial.

$\Leftarrow$: Let $E \in \mu D$ say. Then $(E,J) \in \mu(D, (st\, J) \cap D)$. ○

In the next lemma we consider the $*$-views of a $*$-closed set B;

<u>5.5</u> <u>Lemma</u> For $B \in \mathscr{C}X^*$,

(1) For $E \in ns\mathscr{D}^*$, $^\circ(EB) = {}^\circ E\, {}^\circ B$.

(2) $st(\mathscr{D}^* B) = \mathscr{D}\, {}^\circ B$.

Proof:

(1) Let $D = {}^\circ E$. Since $EB = (E, B \cap E)$ and $D\,{}^\circ B = D\, st\, B = (D, (st\, B) \cap D)$, by 5.1 we need to show that $st(B \cap E) \cap D = (st\, B) \cap D$. This holds since if $x \in D$ with $x \in st\, B$, then moreover $x \in st(E \cap B)$ using that $\mu x \subseteq E$.

(2) By (1) we have $st(\mathscr{D}^* B) \subseteq \mathscr{D}\,{}^\circ B$. And the reverse inclusion holds since for all $D \in \mathscr{D}$, by (1) $D\,{}^\circ B = {}^\circ(D^* B) \in st(\mathscr{D}^* B)$. ○

5.6 Proposition For $D \in \mathscr{D}$, $A \in \mathscr{C}X$, $E \in \mathscr{D}^*$ and $B \in \mathscr{C}X^*$,

$$EB \in \mu\, DA \Leftrightarrow E \in \mu D \text{ and } B \in \mu_D A .$$

Proof:
Both sides imply $E \in \mu D$, so assuming now that $E \in \mu D$ it remains to show that $EB \in \mu\, DA \Leftrightarrow B \in \mu_D A$. By 5.5(1), $EB \in \mu\, DA \Leftrightarrow DA = D\, st\, B \Leftrightarrow A \cap D = (st\, B) \cap D \Leftrightarrow B \in \mu_D A$, as required. ο

The **view map** is the map $\mathscr{D} \times \mathscr{C}X \to \mathscr{V}$ in which $(D,A) \mapsto DA =$ the D-view of A, and we shall denote it explicitly by ν when the need arises. Since $\mathscr{D}, \mathscr{C}X$, and $\mathscr{V}$ are all now equipped with topologies, we may consider the topological aspect of ν. As mentioned in 'Preliminaries', a map $f\colon Z \to Y$ between topological spaces is said to be *perfect* if it is continuous and closed with compact fibres $f^{-1}y$, the nonstandard formulation being that $\forall y \in Y, f^{-1}\mu y = (f^{-1}y)^{\mu}$ (the inclusion '$\supseteq$' expressing continuity, the inclusion '$\subseteq$' expressing the rest). And if f is also surjective, f is an identification (or 'quotient') map (i.e Y has the *finest* topology such that f is continuous), and Y is locally compact iff Z is.

5.7 Proposition The view map is a perfect surjection.
Proof:
We know the map is surjective, and it's continuous by 5.5(1). It now only remains to show that for all $(D,I) \in \mathscr{V}$, $\nu^{-1}\mu(D,I) \subseteq (\nu^{-1}(D,I))^{\mu}$. Suppose then that $(E,B) \in \nu^{-1}\mu(D,I)$, i.e. $(E,B) \in \mathscr{D}^* \times \mathscr{C}X^*$ with $EB \in \mu(D,I)$. Then by 5.5(1), $(D,I) = D\,^{\circ}B = \nu(D, {}^{\circ}B)$ with $(E,B) \in \mu(D, {}^{\circ}B)$, so $(E,B) \in (\nu^{-1}(D,I))^{\mu}$ as required. ο

In particular, continuity of the view map gives that

(1) For $A \in \mathscr{C}X$, the map $\mathscr{D} \to \mathscr{V}$ in which $D \mapsto DA$ is continuous, expressing that "as D varies smoothly so does the D-view of A".
(2) For $D \in \mathscr{D}$, the map $\mathscr{C}X \to \mathscr{V}$ in which $A \mapsto DA$ is continuous, expressing that "as A varies smoothly so does the D-view of A".

Furthermore we have the following characterization of the view topology ;

5.8 Corollary The view topology is the finest topology on $\mathscr{V}$ with respect to which the view map is continuous. ο

And noting that (as $\mathscr{C}X$ is locally compact, being compact) $\mathscr{D} \times \mathscr{C}X$ is locally compact iff $\mathscr{D}$ is locally compact, we have . . .

5.9 Corollary $\mathscr{V}$ is locally compact iff $\mathscr{D}$ is. ο

In particular, as noted in Appendix 8 (just before A8.3), the usual view structure $\mathscr{D}$ on $\mathbb{R}^n$ is locally compact; thus $\mathscr{V}$ is too.

5.10 **Corollary** For $A \in \mathscr{C}X$, the set $\mathscr{D}A$ of views of A is closed in $\mathscr{V}$.

Proof:

$\mathscr{D}A$ is the image of the closed set $\mathscr{D} \times \{A\}$ under the (closed) view map. o

For $A \in \mathscr{C}X$ then, $\mathscr{D}A \in \mathscr{C}\mathscr{V}$. Now if $\mathscr{D}$ is locally compact, so $\mathscr{V}$ is too, we can give $\mathscr{C}\mathscr{V}$ the S-compact topology, and we then obtain a topology on $\mathscr{C}X$ induced by the map $A \mapsto \mathscr{D}A$, in other words giving $B \in \mu A \Leftrightarrow \mathscr{D}^*B \in \mu_s \mathscr{D}A$. The following shows that this is in fact just the view-induced topology again ;

5.11 **Proposition** If $\mathscr{D}$ is locally compact and $\mathscr{C}\mathscr{V}$ has the S-compact topology, then the view-induced topology on $\mathscr{C}X$ is the topology induced by the map $A \mapsto \mathscr{D}A$ of $\mathscr{C}X$ into $\mathscr{C}\mathscr{V}$.

Proof:

We must show that for $A \in \mathscr{C}X$ and $B \in \mathscr{C}X^*$, $B \in \mu_s A \Leftrightarrow st(\mathscr{D}^*B) = \mathscr{D}A$. By 5.5(2) $st(\mathscr{D}^*B) = \mathscr{D}\,{}^\circ B$. So $st(\mathscr{D}^*B) = \mathscr{D}A \Leftrightarrow \mathscr{D}\,{}^\circ B = \mathscr{D}A \Leftrightarrow {}^\circ B = A \Leftrightarrow B \in \mu_s A$, as required. o

Looked at another way, the above essentially says that the map $A \mapsto \mathscr{D}A$ is an embedding of $\mathscr{C}X$ in $\mathscr{C}\mathscr{V}$.

5.12 **Proposition** For $D \in \mathscr{D}$, the set $D\mathscr{C}X = \{(D,I) \mid I$ is a closed-in-D subset of $D\}$ of D-views is homeomorphic to $\mathscr{C}D$ (the set of closed-in-D subsets of D, under the S-compact topology) via the natural map $(D,I) \mapsto I$, and is compact.

Proof:

The map is of course a bijection, and it's moreover a homeomorphism since for $I \in \mathscr{C}D$ and $J \in \mathscr{C}D^*$, $(D^*,J) \in \mu(D,I) \Leftrightarrow st(J \cap D^\mu) = I \Leftrightarrow J \in \mu_s I$ since $st(J \cap D^\mu)$ is the standard part of J in D^*. Since $\mathscr{C}D$ is compact so is $D\mathscr{C}X$ (alternatively note that $D\mathscr{C}X$ is the image of the compact set $\{D\} \times \mathscr{C}X$ under the view map). o

The next result shows that the subview relation $\leq$ is preserved by ${}^\circ$;

5.13 **Proposition** For nearstandard $u_1, u_2 \in \mathscr{V}^*$ with $u_1 \leq u_2$, ${}^\circ u_1 \leq {}^\circ u_2$.

Proof:

Let $u_i = (E_i, J_i) \in \mu v_i$; say $v_i = (D_i, I_i)$. Then $D_1 = subst\, E_1 \subseteq subst\, E_2 = D_2$, so to show that $v_1 \leq v_2$ it remains to show that for $x \in D_1$, $x \in I_1 \Leftrightarrow x \in I_2$, i.e. $x \in D_1 \cap st\, J_1 \Leftrightarrow x \in D_2 \cap st\, J_2$, i.e. $x \in st\, J_1 \Leftrightarrow x \in st\, J_2$. And this holds since, as $J_1 = J_2 \cap E_1$ and $\mu x \subseteq E_1$, $J_1 \cap \mu x = (J_2 \cap E_1) \cap \mu x = J_2 \cap (E_1 \cap \mu x) = J_2 \cap \mu x$. o

Assume now that G is a group of homeomorphisms of X such that, giving G the compact-open topology, G is a *topological* group, i.e. (since we automatically have continuity of composition) such that the inverse map ${}^{-1}$ is continuous, *equivalently* (as is easily shown) such that whenever $g \in G^*$ and $f \in G$ with $g \in \mu f$, $\forall x \in X\ \ g\,\mu x = f\,\mu x$.

As $\mathscr{C}X$ is closed under the elements of G, $(G,\mathscr{D})$ is a similarity view structure on the space X, and we have the following result ;

5.14 Proposition The natural group actions of G on $\mathscr{D}$ and $\mathscr{V}$ are topological, i.e. the evaluation maps $G \times \mathscr{D} \to \mathscr{D}$ and $G \times \mathscr{V} \to \mathscr{V}$ are continuous.

Proof:

Let $g \in G^*$ and $f \in G$ with $g \in \mu f$. So $\forall x \in X \;\; g\,\mu x = f\,\mu x = f\,\mu x$.
(1) Supposing $E \in \mu D$, we need to show $g\,E \in \mu\, f D$. Since $D^{\mu} \subseteq E$, $(f D)^{\mu} = f(D^{\mu}) = g\,(D^{\mu}) \subseteq g\,E$. Since $E \subseteq \overline{D}^{\,\mu}$, $g\,E \subseteq g\,(\overline{D}^{\,\mu}) = f(\overline{D}^{\,\mu}) = (f\overline{D}\,)^{\mu} = \overline{fD}^{\,\mu}$. So $(f D)^{\mu} \subseteq g\,E \subseteq \overline{fD}^{\,\mu}$, i.e. $g\,E \in \mu\, f D$ as required.
(2) Supposing $(E,J) \in \mu\,(D,I)$, we need to show that $g\,(E,J) \in \mu\, f(D,I)$, i.e. $(g\,E, g\,J) \in \mu\,(f D, f I)$. We have $g\,E \in \mu\, f D$ by (1), so it remains to show $\forall y \in f D$, $y \in st\; g\,J \Leftrightarrow y \in f I$, i.e. $\forall x \in D$, $f x \in st\; g\,J \Leftrightarrow f x \in f I$. Now for $x \in D$, $f x \in f I \Leftrightarrow x \in I \Leftrightarrow x \in st\; J$, so the above amounts to showing that $\forall x \in D$, $f x \in st\; g\,J \Leftrightarrow x \in st\; J$. So, let $x \in D$. If $x \in st\; J$ there's $j \in J$ with $j \in \mu x$, and then $g\,j \in g\,J$ with $g\,j \in \mu\, f x$, so $f x \in st\; g\,J$. Conversely, if $f x \in st\; g\,J$ there's $j \in J$ with $g\,j \in \mu\, f x$, and as $g^{-1} \in \mu\, f^{-1}$, $j = g^{-1} g\, j \in \mu\, f^{-1} f x = \mu x$ so $x \in st\; J$. ○

For $g \in G$ then, let $g_{\mathscr{D}}$ denote the homeomorphism of $\mathscr{D}$ induced by g, and let $G_{\mathscr{D}} = \{g_{\mathscr{D}} \mid g \in G\}$. Likewise, let $g_{\mathscr{V}}$ denote the homeomorphism of $\mathscr{V}$ induced by g and let $G_{\mathscr{V}} = \{g_{\mathscr{V}} \mid g \in G\}$. If $\mathscr{D}$ is locally compact, so is $\mathscr{V}$, and we give the groups $G_{\mathscr{D}}$ and $G_{\mathscr{V}}$ of homeomorphisms of $\mathscr{D}$ and $\mathscr{V}$ the compact-open topology, in which case 5.14 says that the natural group actions of G on $\mathscr{D}$ and $\mathscr{V}$ (which are formally the group homomorphisms in which $g \mapsto g_{\mathscr{D}}$ and $g \mapsto g_{\mathscr{V}}$) are continuous.

We now give the set $\mathscr{V}^{\sim}$ of view classes the quotient topology induced from the view topology on $\mathscr{V}$ by the relation of similarity, calling this the <u>view class topology</u>. Since the relation $\sim$ of similarity of views is the equivalence relation induced by the topological action of G on $\mathscr{V}$, it follows from the general result noted in 'Preliminaries' under 'Open Identification Maps', that . . .

5.15 Corollary The quotient map $^{\sim} : \mathscr{V} \to \mathscr{V}^{\sim}$ sending each view to its view class is open. ○

5.16 Note If G acts transitively on $\mathscr{D}$ (i.e. if every two view domains are similar), then $\mathscr{V}^{\sim}$ is compact.

Proof:

Taking $D \in \mathscr{D}$, every view is similar to a D-view, so $\mathscr{V}^{\sim}$ is the image of the set of D-views, which is compact by 5.12, so $\mathscr{V}^{\sim}$ is compact. ○

5.17 Proposition Let $X = \mathbb{R}^n$, $\mathscr{D}$ be the usual view structure on X, and G be a group of similitudes acting transitively on $\mathscr{D}$ with G closed in the group of similitudes of X. Then $\mathscr{V}^{\sim}$ is a compact Hausdorff space.

Proof:

$\mathscr{V}^{\sim}$ is compact by 5.16. To show $\mathscr{V}^{\sim}$ is Hausdorff is equivalent to showing that the relation $\sim$ is closed in $\mathscr{V} \times \mathscr{V}$ (see the section on 'Open Identification Maps' in 'Preliminaries'). So, suppose $u_1, u_2 \in \mathscr{V}^*$ and $v_1, v_2 \in \mathscr{V}$ with $u_1 \in \mu\, v_1$ and $u_2 \in \mu\, v_2$, and $u_1 \sim u_2$; let $g \in G^*$ map u_1 to u_2. It suffices now to show that g is nearstandard in G^*, since if $g \in \mu f$ say, then f will map v_1 to v_2 (for, using 5.14, we'd have $u_2 = g\, u_1 \in \mu\, f\, v_1$, hence since also $u_2 \in \mu\, v_2$ and $\mathscr{V}$ is Hausdorff, $f\, v_1 = v_2$). Let $E_i = dom\ u_i$ and $D_i = dom\ v_i$. Since $E_i \in \mu\, D_i$, it easily follows that E_1 and E_2 are both of finite noninfinitesimal diameter (use that diam : $\mathscr{D} \to (0,\infty)$ is continuous), hence the ratio of g (being diam E_2 / diam E_1) is finite and noninfinitesimal. In addition, there's $x \in X$ such that $g\,x$ is nearstandard (note that as $D_1 \subseteq E_1$, $g\, D_1 \subseteq g\, E_1 = E_2 \subseteq \overline{D_2}^{\,\mu}$). From these facts it follows from A9.2 that g is near a similitude f. And as G is closed in $Sim\ X$, $f \in G$. ○

Finally we show that if $X = \mathbb{R}^n$ and every view domain is convex then the view topology is given by a natural metric as follows. Firstly we define the **silhouette** of a view $v = (D,I)$ to be $\boldsymbol{Sil}\ v = \partial D \cup I$, which, being also $\partial D \cup \overline{I}$ (as $\overline{I} - I \subseteq \partial D$), is compact (as $\partial D = \partial\, \overline{D}$ is compact and $\overline{I}$ is compact, being closed in compact $\overline{D}$) and nonempty (as $\partial D \neq \emptyset$). We thus have what we shall call the **silhouette map** $Sil : \mathscr{V} \to \mathscr{K}X$, and giving $\mathscr{K}X$ the Vietoris topology (which recall is given by the Hausdorff metric) we find that . . .

5.18 Proposition For $X = \mathbb{R}^n$ with every view domain convex, the silhouette map is an embedding.

Proof:

We must show that for $(D,I) \in \mathscr{V}$ and $(E,J) \in \mathscr{V}^*$,

$$(E,J) \in \mu\,(D,I) \iff Sil\,(E,J) \in \mu\ Sil\,(D,I)$$
$$\text{i.e. } \partial E \cup J \in \mu\,(\partial D \cup I).$$

Note that injectivity of Sil will automatically follow due to $\mathscr{K}X$ being Hausdorff.

$\Rightarrow$: As $E \in \mu D$ then $\overline{E} \in \mu\, \overline{D}$ so by continuity of ∂ (proved in A6.4), $\partial\, \overline{E} \in \mu\, \partial\, \overline{D}$, i.e. $\partial E \in \mu\, \partial D$, so $\partial D = st\ \partial E$. We now have

$$\begin{aligned} \partial D \cup I &= \partial D \cup (D \cap st\ J) && \text{as } I = D \cap st\ J \\ &= \partial D \cup st\ J && \text{as } st\ J - \partial D \subseteq \overline{D} - \partial D = D \\ &= st\ \partial E \cup st\ J && \text{as } \partial D = st\ \partial E \\ &= st\,(\partial E \cup J). \end{aligned}$$

Since also $\partial E \cup J \subseteq \overline{E} \subseteq \overline{D}^{\,\mu} \subseteq ns\ X^*$ we have $\partial E \cup J \in \mu\,(\partial D \cup I)$ as required.

$\Leftarrow$: By continuity of $conv : \mathscr{K}X \to \mathscr{K}X$ (see A8.5), $conv(\partial E \cup J) \in \mu\, conv(\partial D \cup I)$, i.e. $\overline{E} \in \mu\, \overline{D}$, hence $E \in \mu D$ (and also $\partial D = st\, \partial E$ as proved above in '$\Rightarrow$'). It now remains to show that $I = (st\, J) \cap D$. We know that $\partial D \cup I = st(\partial E \cup J)$, so

$$\begin{aligned} I &= st(\partial E \cup J) - \partial D \\ &= (st\, \partial E \cup st\, J) - \partial D \\ &= (\partial D \cup st\, J) - \partial D \\ &= st\, J - \partial D \\ &= (st\, J) \cap D \qquad \text{as } st\, J \subsetneq st\, \overline{E} = \overline{D}. \end{aligned}$$

o

Thus, defining the <u>view metric</u> to be the metric d on $\mathscr{V}$ induced from the Hausdorff metric h by the silhouette map, i.e. given by $d(u, v) = h(Sil\, u, Sil\, v)$, we have the following result;

<u>5.19</u> <u>Corollary</u> For $X = \mathbb{R}^n$ with every view domain convex, the view topology is given by the view metric. o

Note that if we'd only asked that the view domains be *connected*, the above work would break down because the silhouette map would not necessarily be injective. For example, let $X = \mathbb{R}^2$, D_1 be the open unit ball $(0)_1$, $I_1 = [0]_{1/2}$, $D_2 = D_1 - I_1$, and $I_2 = \emptyset$. Then $(D_1, I_1) \neq (D_2, I_2)$ yet their silhouettes are both $\partial D_1 \cup \partial I_1$.

The view metric can also be used to naturally metrize the compact Hausdorff space $\mathscr{V}^{\sim}$ of view classes in the case of the usual similarity view structure on $\mathbb{R}^n$; see note 7.11 for details.

6. A Definition of 'Visually Fractal'

Throughout let $X = \mathbb{R}^n$, $\mathscr{D}$ be the usual view structure on X, and $\mathscr{V} = \mathscr{D}\mathscr{C}X$. G will denote the group of rotation-free direct similitudes, namely generated by the translations and dilations, and the relation of similarity among views with respect to G will be denoted by $\equiv$. We shall call the resulting view classes oriented view classes, these being (for $n \geq 2$) distinct from the more usual type of view class arising from the *whole* group of direct similitudes. Intuitively, oriented view classes model the idea of images coming from a remote camera whose orientation is *constant*. For all $D_1, D_2 \in \mathscr{D}$ there's a *unique* $g \in G$ mapping D_1 to D_2, so in particular G acts transitively on $\mathscr{D}$, and since G is also closed in *Sim X* we know by 5.17 that the space $\mathscr{V}^{\equiv}$ of oriented view classes is a compact Hausdorff space.

For $x \in X$ and $r > 0$ let $(x)_r$ denote the open ball (i.e. view domain) with centre x and radius r. For $D \in \mathscr{D}$ the unique $g \in G$ mapping D to the unit open ball $(0)_1$ is said to normalize D. A $(0)_1$-view is called a normal view, and for a view v the map g normalizing $dom\ v$ is said to normalize v, being the unique $g \in G$ mapping v to a normal view; we shall call $g\,v$ the normalization of v, denoted by $\boldsymbol{norm}\ v$. The fact that each view is similar to a unique normal view says that each oriented view class V contains a unique normal view, i.e. has a unique normal realization, and we'll denote it by $\boldsymbol{norm}\ V$. Denoting the set of normal views by $\boldsymbol{norm}\ \mathscr{V}$, then . . .

6.1 Proposition $\mathscr{V}^{\equiv} \cong norm\ \mathscr{V}$ via the map $V \mapsto norm\ V$.

Proof:
The inverse of the bijection in question is $^{\equiv} : norm\ \mathscr{V} \to \mathscr{V}^{\equiv}$, which we already know is continuous. Since $norm\ \mathscr{V}$ is (by 5.12) a compact Hausdorff space, the map $^{\equiv}$ is hence a homeomorphism, hence so is its inverse. ○

We thus have the option of using normal views in place of oriented view classes, replacing use of $v^{\equiv}$ by that of $norm\ v$, which for some purposes is more convenient. The quotient map $^{\equiv} : \mathscr{V} \to \mathscr{V}^{\equiv}$ is replaced by the view normalization map $norm : \mathscr{V} \to norm\ \mathscr{V}$, noting that in general $norm\ v = norm\,(v^{\equiv})$. The latter map thus has the properties of the former, and in particular . . .

6.2 Corollary The view normalization map is an open identification map. ○

Bear in mind that by 5.12 with $D = (0)_1$, $norm\ \mathscr{V}$ is naturally homeomorphic to $\mathscr{C}(0)_1$ with the S-compact topology, $v \in norm\ \mathscr{V}$ corresponding to its object part $ob\ v$. In addition we know from 5.19 that the topology involved is given by the view metric. We can give an independent short proof of this as follows. First note that for $u_1, u_2 \in (norm\ \mathscr{V})^*$, $d(u_1, u_2) \approx 0 \Leftrightarrow st_{(0)_1}\, ob\ u_1 = st_{(0)_1}\, ob\ u_2$. So for $u \in (norm\ \mathscr{V})^*$ and $v \in norm\ \mathscr{V}$,

$$d(u, v^*) \approx 0 \Leftrightarrow st_{(0)_1} ob\, u = st_{(0)_1} ob\, v^*$$
$$\Leftrightarrow st_{(0)_1} ob\, u = ob\, v$$
$$\Leftrightarrow u \in \mu v \text{, as required.}$$

This metric on the normal views makes an implicit appearance in some work of Tan Lei on the Mandelbrot set; see note 7.16 for further details.

For $A \in \mathscr{C}X$ and $x \in X$, an **x-view of A** is a view of A whose domain is a ball centred at x, i.e. a view of the form $(x)_r A$. We now come to the key definitions in this section. Letting $x \in A$, suppose that as we zoom in on x, what we see settles down towards a limit; the *formal* expression of this is that $norm\,(x)_r A$ converges in $norm\,\mathscr{V}$ as r decreases to 0, equivalently that $((x)_r A)^{\equiv}$ converges in $\mathscr{V}^{\equiv}$ as r decreases to 0. If this occurs we'll say A is **visually convergent at x**; otherwise A is **visually fractal at x**, which is one interpretation of the idea of A having 'detail at all levels of magnification' at x. A is thus divided into two parts which we'll call its **convergent part** and **fractal part**, consisting respectively of the points at which A is visually convergent or fractal, and denoted by $\boldsymbol{cv}\,A$ and $\boldsymbol{fr}\,A$. Bearing in mind that $norm\,\mathscr{V}$ is a compact Hausdorff space, the nonstandard formulation that A be visually convergent at x is that $^{\circ}\,norm\,((x)_r A^*)$ be the same for all infinitesimal r. If this holds, the limit in $norm\,\mathscr{V}$ will be called the **limit view of A at x**, denoted by $\boldsymbol{lim}_A\, x$. We'll say A is **visually convergent** if it is visually convergent everywhere, equivalently if it is visually convergent at its boundary points (since trivially it's visually convergent at any interior point). Likewise A is **visually fractal** if it is visually fractal everywhere, which implies that A is residual. To avoid constant repetition in what is now an established context, in the following work we shall generally drop the adjective 'visually' on the above definitions, except in the formal statement of results. Some examples of convergent sets in $X = \mathbb{R}^2$ are illustrated below ;

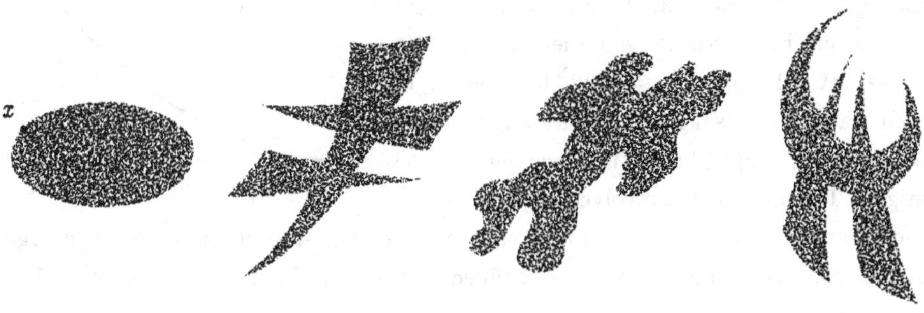

Note that in the case of the first example, the limit view at any boundary point is a view of a half-plane just touching the origin. This is illustrated in the sketch on the next page, which shows the normalizations of sucessive views of A at a boundary point x, converging to the limit view as we zoom in on x ;

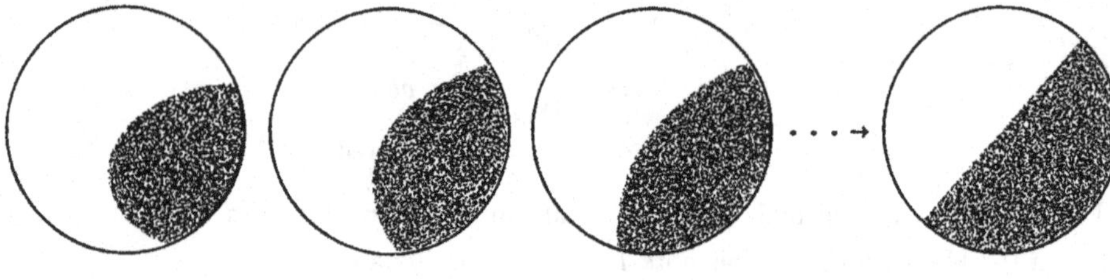

These pictures also suggest the proof of the following result;

<u>**6.3 Proposition**</u> Every convex closed set is visually convergent.

Proof:

Let A be a convex closed set, and take any $x \in A$. For $r > 0$ let $I_r = ob\ norm\,(x)_r\, A$. As A is convex about x, I_r increases as r decreases. So using A7.4(2), as r decreases to 0, I_r converges in $\mathscr{C}(0)_1$ (namely to the closure of $\bigcup_r I_r = \bigcup_n I_{1/n}$ in $(0)_1$), in other words (recalling the way in which $norm\,\mathscr{V}$ is homeomorphic to $\mathscr{C}(0)_1$) $norm\,(x)_r\, A$ converges in $norm\,\mathscr{V}$ as required. ○

Indeed, by A7.4(2) a set A is convergent at any point x for which $ob\ norm\,(x)_r\, A$ is eventually increasing or eventually decreasing as r decreases to 0.

Later we shall provide a sufficient condition for certain of Hutchinson's invariant sets K_F to be fractal. In the meanwhile we note one simple example of a set being fractal at a point. Namely, let $X = \mathbb{R}$ and $A = \{0\} \cup \{1/2^n \mid n \geq 0\}$; then A is fractal at 0. Indeed, A is 'visually periodic' at 0 in a sense made precise later on, and this will *imply* fractality at the point in question. Note in the present example that A is convergent at all points $1/2^n$ (a set being convergent at any isolated point), showing that the convergent part of a set A need not be closed in A. On the other hand, the example on the right in $\mathbb{R}^2$ shows that the *fractal* part of a set A need not be closed in A either. Here, A is fractal at all points $a_n = (1/2^n,0)$ (the set of points above a_n being a vertically aligned copy of $\{a_m \mid m \geq 0\}$), but A is convergent at the origin. The upper curve involved is tangential to the lower line at the origin, and it should be noted that A would have been convergent at the origin *regardless* of how we had 'filled in' the space between this curve and the lower line.

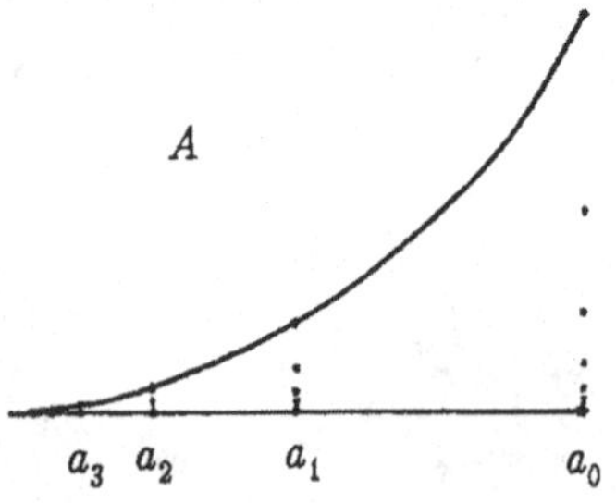

Despite the example in $X = \mathbb{R}$ of $A = \{0\} \cup \{1/2^n \mid n \geq 0\}$ being fractal at 0, note that the set $B = \{0\} \cup \{1/n \mid n \geq 1\}$ is actually *convergent* at 0 (despite the existence of a homeomorphism of X mapping A to B), the object part of the limit view being [0,1). We prove this as follows, letting $b_n = 1/n$. Take any infinitesimal $r > 0$, and let $n \in \mathbb{N}^*$ with $b_n < r \leq b_{n-1}$. Letting g be multiplication by $1/r$, g normalizes $u = (0)_r\, B^*$, and *ob norm* $u = \{0\} \cup \{g\, b_m \mid m \geq n\}$. Now for all $m \geq n$ we have $g\, b_{m+1} \approx g\, b_m$ noting that $g\, b_m - g\, b_{m+1} = (b_m - b_{m+1})/r \leq n\,(b_m - b_{m+1}) = n/(m\,(m+1)) \leq n/(n^2) = 1/n \approx 0$. Also, $g\, b_n \approx 1$. It follows that the standard part of *ob norm* u in $(0)_1$ is [0,1) as required. This example also shows that we can have a set B which is *residual*, yet having a limit view whose object part is *not* residual in $(0)_1$; indeed, it can have a limit view which is *full*.

A limit view is a view of the form $lim_A\, x$ for some $x \in A \in \mathscr{C}X$, and we'll provide a complete description of these shortly, facilitating later work. First, for $\lambda \in (0,1)$ we define a zoom operation z_λ on *norm* $\mathscr{V}$ by

$$z_\lambda\,((0)_1\,, I\,) = norm\,((0)_\lambda\,, I \cap (0)_\lambda\,).$$

It is easily seen that for $A \in \mathscr{C}X$, $z_\lambda\,(norm\,(x)_r\,A\,) = \text{norm}\,(x)_{\lambda r}\,A$. In addition, . . .

6.4 Lemma $\forall \lambda \in (0,1)$ z_λ is continuous.

Proof:

Let $v \in norm\,\mathscr{V}$ and $u \in (norm\,\mathscr{V})^*$ with $u \in \mu v$, and let $I = ob\, v$ and $J = ob\, u$. Then for $x \in (0)_\lambda$, $x \in st\,(J \cap (0)_\lambda{}^*) \Leftrightarrow x \in st\, J \Leftrightarrow x \in I$; so $st\,(J \cap (0)_\lambda{}^*) \cap (0)_\lambda = I \cap (0)_\lambda$, which says that $((0)_\lambda{}^*, J \cap (0)_\lambda{}^*) \in \mu\,((0)_\lambda\,, I \cap (0)_\lambda\,)$, so by continuity of the normalization map (part of 6.2) $norm\,((0)_\lambda{}^*, J \cap (0)_\lambda{}^*) \in \mu\; norm\,((0)_\lambda\,, I \cap (0)_\lambda\,)$, i.e. $z_\lambda u \in \mu\, z_\lambda v$. ○

We'll say a view $v = ((x)_r\,, I\,)$ is radiant if $I = \{x\} \cup \{\,[x,y) \mid y \in Y\}$ for some $Y \subseteq \partial\,(x)_r$. In this case, Y is unique, easily shown to be $\overline{I} - I$, and we'll call Y the radiant set of v, denoted by ***rad v***, this being a closed subset of the sphere $\partial\,(x)_r$ (the unit sphere S^{n-1} in the case v is normal). Some examples of radiant views in $\mathbb{R}^2$ are illustrated below;

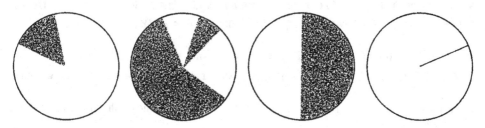

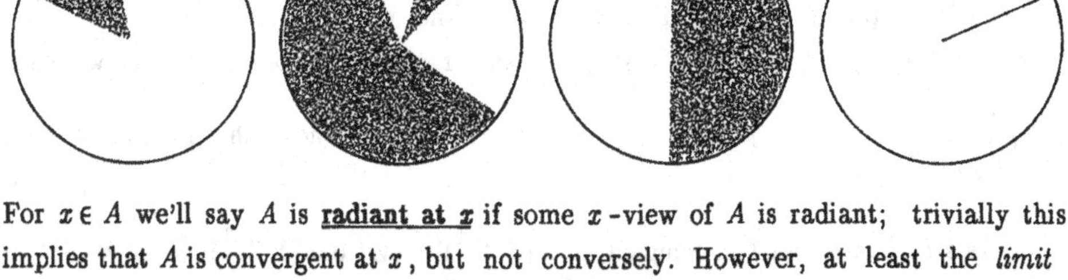

For $x \in A$ we'll say A is radiant at x if some x-view of A is radiant; trivially this implies that A is convergent at x, but not conversely. However, at least the *limit* view of A at x is radiant; for as we now show, the limit views are precisely the radiant normal views. For closed $Y \subseteq S^{n-1}$ let $IY = \{0\} \cup \{\,[0,y) \mid y \in Y\}$.

6.5 Proposition For a normal view v the following are equivalent;

(1) v is a limit view.

(2) v is radiant.

(3) v is nonempty and invariant under every zoom operation z_λ.

Proof:

(1) $\Rightarrow$ (3) : Say $v = \mathit{lim}_A\, x$, and for $r > 0$ let $u_r = \mathit{norm}\,(x)_r\, A$. Naturally v is nonempty since $0 \in \mathit{ob}\, v$. And for $\lambda \in (0,1)$, since v is the limit of u_r as r decreases to 0 then by continuity of z_λ, $z_\lambda v$ is the limit of $z_\lambda u_r = u_{\lambda r}$ as r decreases to 0 ; but this is of course just the same limit, so $z_\lambda v = v$, i.e. v is invariant under z_λ.

(3) $\Rightarrow$ (2) : Let $I = \mathit{ob}\, v$. If $I - \{0\} = \emptyset$ then $I = \{x\}$ so v is radiant. Now assume $I - \{0\} \neq \emptyset$. From (3) it easily follows that for all $p \in I - \{0\}$, $(0,y) \subseteq I$ where y is the element of S^{n-1} such that $p \in (0,y)$; moreover $[0,y) \subseteq I$ as I is closed in $(0)_1$. So I is a nonempty union of sets of the form $[0,y)$ with $y \in S^{n-1}$, hence v is radiant.

(2) $\Rightarrow$ (1) : Then letting $Y = \mathit{rad}\, v$ and $A = \overline{IY}$ (which is a closed subset of $[0]_1$), A is radiant at 0, and $v = (0)_1 A = \mathit{lim}_A\, 0$, so v is a limit view. ○

Using this, we thus have the following homeomorphism between the space $\mathbf{lim}\,\mathscr{V}$ of limit views and $\mathscr{C}(S^{n-1})$, defining for $Y \in \mathscr{C}(S^{n-1})$ that $\mathbf{lim}\, Y = ((0)_1, IY)$;

6.6 Proposition Giving $\mathscr{C}(S^{n-1})$ the Vietoris topology, $\mathit{lim}\,\mathscr{V} \cong \mathscr{C}(S^{n-1})$ via the mutually inverse maps $v \mapsto \mathit{rad}\, v$ and $Y \mapsto \mathit{lim}\, Y$.

Proof:

The fact that these maps *are* mutually inverse bijections is straightforward. We now show that the second is a homeomorphism, which (recalling how $\mathit{norm}\,\mathscr{V} \cong \mathscr{C}(0)_1$, and denoting monads in the latter space by μ_s, given by $B \in \mu_s A \Leftrightarrow st_{(0)_1} B = A$) reduces to showing that for $Y \in \mathscr{C}(S^{n-1})$ and $Z \in \mathscr{C}(S^{n-1})^*$, $Z \in \mu Y \Leftrightarrow IZ \in \mu_s\, IY$. This is proved as follows, first of all noting that for $y \in S^{n-1}$ and $z \in (S^{n-1})^*$, $z \in \mu y \Leftrightarrow [0,z) \in \mu_s\, [0,y)$.

$\Rightarrow$: $Z \in \mu Y$ means that $\forall y \in Y\ \exists z \in Z\ z \in \mu y$ and $\forall z \in Z\ \exists y \in Y\ z \in \mu y$. By the above note it easily follows that $st_{(0)_1} IZ = IY$, i.e. $IZ \in \mu_s\, IY$.

$\Leftarrow$: We need to prove that $\forall y \in Y\ \exists z \in Z\ z \in \mu y$ and $\forall z \in Z\ \exists y \in Y\ z \in \mu y$. Taking $y \in Y$, $y/2 \in [0,y) \subseteq IY = st_{(0)_1} IZ$ so for some $z \in Z$ and $\alpha \in [0,z)$ we have $\alpha \in \mu\,(y/2)$, which easily gives $z \in \mu y$. A similar argument shows that for $z \in Z$ there's $y \in Y$ with $z \in \mu y$. ○

As a corollary, $\mathit{lim}\,\mathscr{V}$ is compact as $\mathscr{C}(S^{n-1}) = \{\emptyset\} \cup \mathscr{K}(S^{n-1})$ with $\mathscr{K}(S^{n-1})$ compact. Note that $\emptyset$, isolated in $\mathscr{C}(S^{n-1})$, corresponds to the limit view with object part $\{0\}$, which is the limit view of a set at any of its isolated points (indeed, for $x \in A$,

x is isolated in A iff A is convergent at x with $rad\ lim_A x = \emptyset$). At the other end of the scale, S^{n-1} corresponds to the *full* limit view, which is the limit view of a set at any of its interior points.

We now give a sufficient condition that a residual invariant set $K = K_F$ be fractal ;

6.7 Proposition Let F be a finite set of contractive direct similitudes which satisfies the compact set condition with some A such that $\underline{A}\,K$ is not embedded (by a direct similitude) in any limit view, and $|K| \geq 2$. Then K is visually fractal.

Proof:

Let $x \in K$. Taking infinite n, let $f \in F^n$ with $x \in f(K^*)$, and let $\epsilon = r_f$ diam A, which is infinitesimal. Note that $f(A^*) \subseteq [x]_\epsilon$ as $x \in f(A^*)$ with diam $f(A^*) = r_f$ diam $A^* = \epsilon$. Where $g \in G^*$ normalizes $(x)_\epsilon$, we have $r_g = 1/\epsilon$ so $r_{g \circ f} = r_g r_f = 1/$ diam A, which in particular is a finite noninfinitesimal, and since $g \circ f$ also maps a bounded point to a bounded point, by A9.2 it's near some similitude h.

Letting $D = \underline{A}$, we now show that h embeds DK in $^\circ(norm\,(x)_\epsilon\, K^*)$; for it will then follow by our suppositions that the latter is not a limit view, so K will be fractal at x as required. By transfer of 4.1(1), and the fact that $f(D^*) \subseteq (x)_\epsilon$, f embeds $(DK)^*$ in $(x)_\epsilon\, K^*$, i.e. $f\,(DK)^* \leq (x)_\epsilon\, K^*$, hence $g\,f\,(DK)^* \leq g\,(x)_\epsilon\, K^*$, i.e. $(g \circ f)(DK)^* \leq norm\,(x)_\epsilon\, K^*$, so by 5.13 $^\circ((g \circ f)(DK)^*) \leq {}^\circ(norm\,(x)_\epsilon\, K^*)$. And since $(g \circ f) \in \mu h$, then by 5.14 $^\circ((g \circ f)(DK)^*) = h\,(DK)$. So $h\,(DK) \leq {}^\circ(norm\,(x)_\epsilon\, K^*)$ as required. ○

One easily seen sufficient condition that $\underline{A}\,K$ not be embedded in any limit view is that $K \cap \underline{A}$ (the object part of $\underline{A}\,K$) contain at least two points and expand no line segment $[x,y]$; and since this condition in particular holds if K is totally disconnected and $K \cap \underline{A} \neq \emptyset$ (hence moreover $|K \cap \underline{A}| \geq 2$ using 10.7 of Chapter 3), note that . . .

6.8 Corollary Let F be a finite set of contractive direct similitudes which satisfies the compact set condition with some A such that $K \cap \underline{A} \neq \emptyset$, and suppose K is totally disconnected. Then K is visually fractal. ○

In particular the Cantor set in $\mathbb{R}$ is fractal. In *practice*, with $X = \mathbb{R}^2$ a computer-generated picture of K will quickly suggest whether or not $\underline{A}\,K$ can be embedded in a limit view, and more often than not the answer is likely to be 'no' as required, since views embeddable in a limit view are rather special. For example, the Sierpinski Gasket is clearly fractal, by looking at $\underline{A}\,K$ where A is the triangle involved. A *formal* proof is provided by the following strong result, ***aff** K* denoting the affine span of K;

6.9 Proposition Let $X = \mathbb{R}^2$ and F be a finite set of contractive direct similitudes which satisfies the compact set condition with some $A \neq K$ such that $K \cap \underline{A} \neq \emptyset$ and $aff\,K = X$. Then K is visually fractal.

Proof:

Firstly note that K is residual by 7.13 of Chapter 3. Letting $D = \underline{A}$ we show that DK is not embedded in any limit view, which by 6.7 will imply that K is fractal. Suppose for a contradiction that there *is* a direct similitude h embedding DK in a limit view v. This is illustrated on the right. Using that K is residual and v is radiant, it follows that for all $x \in ob\,DK - h^{-1}0$ there's a unique line L_x through x such that some neighbourhood of x in L_x is a subset of $ob\,DK$. Now $ob\,DK = K \cap D$ is not a subset of a line (otherwise, as

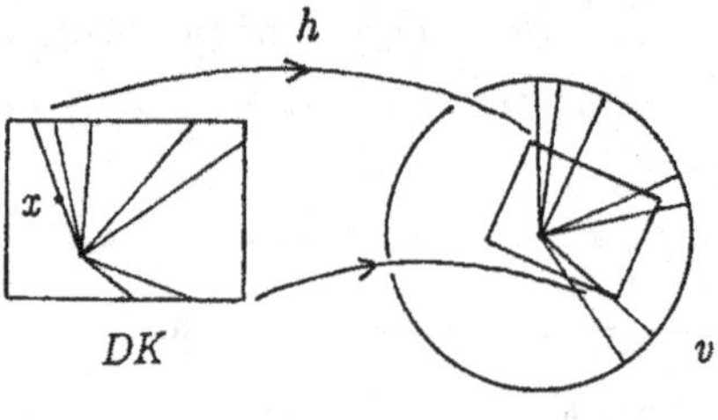

$K \cap D$ is dense in K by 6.3 of Chapter 3, K would be a subset of a line, i.e. $aff\,K \neq X$, contradicting one of our assumptions) so (since in particular $|ob\,DK| \geq 2$) let $x \in ob\,DK - h^{-1}0$, and then let $y \in ob\,DK$ with $y \notin L_x$. Note that L_y is defined since $h^{-1}0 \in L_x$. Taking a $*$-neighbourhood N of x with $N \subsetneq \mu x$, by transfer of result 6.1(2) of Chapter 3 there exists some $f \in (Semigroup\ F)^*$ with $f(D^*) \subsetneq N$, and by 4.1(1) f embeds $(DK)^*$ in $(DK)^*$. As f is a $*$-similitude the angle

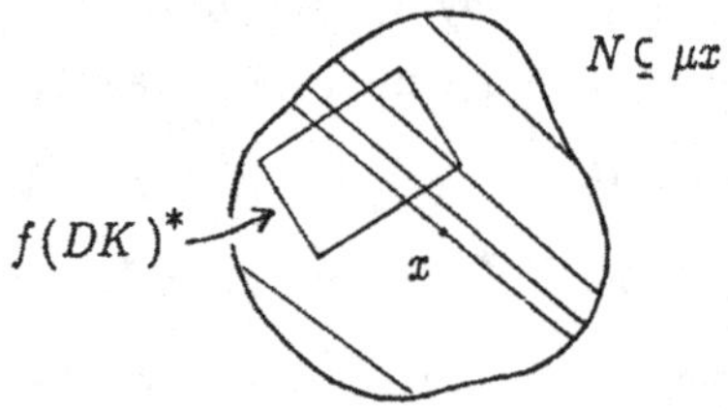

between $f(L_x{}^*)$ and $f(L_y{}^*)$ is the same as that between $L_x{}^*$ and $L_y{}^*$ (namely that between L_x and L_y). But by transfer of a standard fact, $f(L_x{}^*) = L^*{}_{fx}$ and $f(L_y{}^*) = L^*{}_{fy}$, the angle between which is *infinitesimal* using that $fx \approx fy$ (so $h\,fx \approx h\,fy$). We thus have a contradiction as required. ○

Changing direction, we now ask whether for $A \in \mathscr{C}X$ and $x \in \partial A$ it is generally true that A is fractal at x iff ∂A is fractal at x. Of course, this is only of interest for non-residual sets. The answer is that the forward implication holds whilst the reverse fails. As an example of this failure in $\mathbb{R}$, let $a_n = 1/2^n$ for $n \geq 0$, and let $A = \{0\} \cup \bigcup\{\, [a_n + \epsilon_n, a_{n-1}] \mid n \geq 1 \,\}$ where $\epsilon_n = (1/n)\, d(a_n, a_{n-1}) = (1/n) a_n$. Then $\partial A = \{0\} \cup \{a_n \mid n \geq 0\} \cup \{a_n + \epsilon_n \mid n \geq 1\}$ is fractal at 0 but A is not, all due to the fact that $\epsilon_n / a_n = 1/n$ converges to 0 as n tends to ∞. We now however give the proof of the *forward* implication above ;

6.10 Proposition For $A \in \mathscr{C}X$ and $x \in \partial A$, if A is visually fractal at x so is ∂A.

Proof:

We prove the contrapositive; assuming that ∂A is convergent at x we now show that

A is convergent at x too. Let $Y = rad\ lim_{\partial A}\ x$.

For $z \in S^{n-1} - Y$, the set $M_z = \mu x \cap \bigcup\{ (x,w] \mid w \in \mu_{(S^{n-1})} z \}$ is either a subset of A^* or is disjoint from A^*. For, suppose $\alpha \in M_z \cap A^*$. Then taking any other $\beta \in M_z$ we have $\beta \in A^*$; for if $\beta \notin A^*$ then by transfer of a standard fact the line segment $[\alpha, \beta]$ must cross the boundary of A^* at some point γ, but this would give the contradiction that $z \in Y$ (for where $r = 2\ d(x,\gamma)$ and $g \in G^*$ normalizes $(x)_r$, then $z/2 = {}^\circ(g\ \gamma) \in {}^\circ(ob\ norm\,(x)_r\ \partial A^*) = ob\ lim_{\partial A}\ x$ giving $z \in Y$ as $lim_{\partial A}\ x$ is radiant).

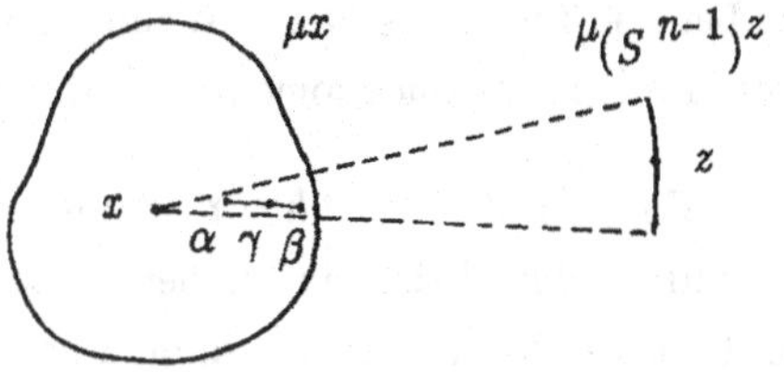

Letting $Z = \{ z \in S^{n-1} - Y \mid M_z \subseteq A^* \}$ then, it is easily seen that for any infinitesimal r, ${}^\circ(norm\,(x)_r\ A^*)$ is radiant, with radiant set $Y \cup Z$ (Y coming from the fact that $\partial A \subseteq A$). So A is convergent at x as required. ○

We continue with a few further results concerning boundaries. For a view $v = (D,I)$ we define $\partial v = (D, \partial_D I)$, where ∂_D denotes 'boundary in D' (note that $\partial_D I = D \cap \partial D$ since D is open). ∂v is a view with the same domain as v, and the following result applies to *any* topological view space ;

6.11 Proposition For $D \in \mathscr{D}$ and $A \in \mathscr{C}X$, $\partial\, DA = D\, \partial A$.

Proof:
We must show that $\partial_D (D \cap A) = D \cap \partial A$, i.e. $D \cap \partial(D \cap A) = D \cap \partial A$, i.e. for $x \in D$, $x \in \partial(D \cap A) \Leftrightarrow x \in \partial A$. Using that $\mu x \subseteq D^*$, this is proved as follows;

$$\begin{aligned} x \in \partial(D \cap A) &\Leftrightarrow \exists \alpha,\beta \in \mu x \text{ with } \alpha \in (D \cap A)^* \text{ and } \beta \notin (D \cap A)^* \\ &\Leftrightarrow \exists \alpha,\beta \in \mu x \text{ with } \alpha \in D^* \cap A^* \text{ and } \beta \notin D^* \cap A^* \\ &\Leftrightarrow \exists \alpha,\beta \in \mu x \text{ with } \alpha \in A^* \text{ and } \beta \notin A^* \\ &\Leftrightarrow x \in \partial A . \end{aligned}$$
○

Note that the map $\partial : \mathscr{V} \to \mathscr{V}$ is not continuous. For example, in $X = \mathbb{R}$, let $v \in norm\ \mathscr{V}$ and $u \in (norm\ \mathscr{V})^*$ with v full and $ob\ u = (0)_1{}^* - (0)_\epsilon$ for some infinitesimal $\epsilon > 0$. Then $u \in \mu v$ but $\partial u \notin \mu\ \partial v$; indeed, ∂u is near the normal view with object part $\{0\}$.

The next result describes ∂v in the case where v is a non-full limit view ;

6.12 Proposition For $Y \in \mathscr{C}(S^{n-1})$ with $Y \neq S^{n-1}$, $\partial\ lim\ Y = lim\ \partial_{(S^{n-1})} Y$.

Proof:
Let $Z = \partial_{(S^{n-1})} Y$. We must show $\partial_{(0)_1} IY = IZ$, i.e. $\forall x \in (0)_1$, $x \in \partial\, IY \Leftrightarrow x \in IZ$.

Both sides hold for $x = 0$ (using $Y \neq S^{n-1}$ to show $0 \in \partial IY$), so now take $x \neq 0$.
$\Rightarrow$: Let $\beta \in \mu x$ with $\beta \notin IY^*$, and let $\alpha \in (S^{n-1})^*$ with $\beta \in (0,\alpha)$. Since $\beta \notin IY^*$, $\alpha \notin Y^*$. Now $x \in (0)_1 \cap \partial IY = \partial_{(0)_1} IY \subseteq IY$ so let $y \in Y$ with $x \in (0,y)$. Then $\alpha \in \mu y$ so $y \in Z$, hence $x \in IZ$ as required.
$\Leftarrow$: Let $z \in Z$ with $x \in (0,z)$. Where $\alpha \in (S^{n-1})^* - Y^*$ with $\alpha \in \mu z$, $(0,\alpha)$ is disjoint from IY^*, hence since some $\beta \in (0,\alpha)$ is near x we have $x \in \partial IY$. o

For $Y = S^{n-1}$, $\partial\, lim\, Y$ is the *empty* normal view. Although we have not formally included this view under the term 'limit view', note that it does arise as such, in the sense that if we zoom in on a point $x \notin A$ we eventually see only an empty view; $norm\,(x)_r A$ is the empty normal view for all sufficiently small r. By 6.12, the set consisting of the limit views and the empty normal view is closed under ∂.

We'll say a view $v = (D,I)$ is <u>residual</u> if I is residual in D, equivalently $\partial_D I = I$, equivalently $\partial v = v$. Then . . .

<u>6.13</u> <u>Note</u> For a limit view v, v is residual $\Leftrightarrow$ *rad* v is residual in S^{n-1}.
Proof:
Equivalently we must show that for $Y \in \mathscr{C}(S^{n-1})$, *lim* Y is residual $\Leftrightarrow$ Y is residual in S^{n-1}. It's true for $Y = S^{n-1}$ as both sides fail. For $Y \neq S^{n-1}$, using 6.12, *lim* Y is residual $\Leftrightarrow$ $\partial\, lim\, Y = lim\, Y \Leftrightarrow lim\, \partial_{(S^{n-1})} Y = lim\, Y \Leftrightarrow \partial_{(S^{n-1})} Y = Y \Leftrightarrow Y$ is residual in S^{n-1}. o

Finally, we describe the notion of visual periodicity mentioned in passing earlier on. For $A \in \mathscr{C}X$ and $x \in A$ we'll say A is <u>visually periodic</u> at x if for some $r_1 > r_2 > 0$, $(x)_{r_2} A \equiv (x)_{r_1} A$ and $\forall \delta \in (r_2, r_1)$ $(x)_\delta A \not\equiv (x)_{r_1} A$. If this holds then r_1/r_2 is the same for *all* such pairs (r_1, r_2) (*Proof*: take any such pair (δ_1, δ_2) and assume $\delta_1 \le r_1$. Now where $\delta_1 = \lambda r_1$, $(\lambda r_1, \lambda r_2)$ is clearly such a pair too, hence we must have $\delta_2 = \lambda r_2$, thus $\delta_1/\delta_2 = r_1/r_2$), and we define the <u>period</u> of A at x to be $ln\,(r_1/r_2) = ln\, r_1 - ln\, r_2$. The reason for this definition of the period (rather than r_1/r_2 say) is as follows. Imagine zooming in on x at a constant rate compared to the existing size of the view domain. More precisely, suppose that where at time t our view domain is $(x)_{r(t)}$, we have $r(0) = 1$ and $\dot{r}(t) = -r(t)$, so $r(t) = e^{-t}$. Then taking $r_1 = r(t_1)$ and $r_2 = r(t_2)$ we have $ln\, r_1 - ln\, r_2 = t_2 - t_1$, the time taken between the appearance of successive identical normalized views (once we've zoomed in far enough). Visual periodicity of A at x thus equivalently states that the map $[0,\infty) \to norm\, \mathscr{V}$ in which $t \mapsto norm\,(x)_{r(t)} A$ is eventually periodic (the period then being that of A at x), and it is immediately apparent that in such a case $norm\,(x)_{r(t)} A$ does not converge as t increases to ∞ (i.e. as $r(t)$ decreases to 0), so A is fractal at x;

<u>6.14</u> <u>Note</u> A is visually periodic at x $\Rightarrow$ A is visually fractal at x. o

7. Notes, Questions, and Suggestions for Further Work

The notes below are grouped under the sections to which they are most relevant, except for the final 'miscellaneous' group. Several of the notes simply point out various natural view-related definitions worth bearing in mind for further work. Throughout, assume unless otherwise implied that we're dealing with $X = \mathbb{R}^n$ with the usual similarity view structure, A and B denoting elements of the 'object set' $\mathscr{C}X$, and the similarity relation being denoted by $\sim$. The symbol $\equiv$ will be reserved for denoting the relation of similarity arising from the *rotation-free* direct similitudes, as described in Section 6. Despite the convention of $X = \mathbb{R}^n$, note that several of the points below (or in some cases, suitable generalizations) make sense for *any* topological similarity view space, or, even more generally, any similarity view space.

Section 1 : Views and Similarities

7.1 Although the term 'view' we have used is quite apt in the case of $\mathbb{R}^2$ (thinking of looking down on a plane and obtaining views of a set), its meaning differs *significantly* from everyday usage in the case of $\mathbb{R}^3$ (not that we *intended* to model the latter usage). Here, our 'views' are as if obtained not from a *normal* camera but from one which can 'see through' structure, taking a whole 'sample' of an area and the object-structure within it. However, if we 'looked down' on $\mathbb{R}^3$ from within $\mathbb{R}^4$, the term 'view' would be appropriate in an analogous way to the earlier $\mathbb{R}^2$ case. The main point here as regards interpretations is that one should be aware that there are two things involved; the space one is studying and the vantage point one is studying it *from*.

7.2 Recall that an x-view is a view whose domain is a ball centred at x. For A we define the equivalence relation $\overset{A}{\sim}$ in A by $x \overset{A}{\sim} y \Leftrightarrow$ some x-view of A is similar to a y-view of A, saying A is **similar at x and y**. The equivalence classes are called the **similarity classes** of A, and should be worth investigating. For example, consider the similarity classes of the Sierpinski Gasket, or other invariant sets K_F. Generalizing $\overset{A}{\sim}$, for $a \in A$ and $b \in B$ we may say A and B are **similar at a and b** if some a-view of A is similar to a b-view of B. For a weakening of this relation see 7.16.

7.3 Does there exist a set A consistent with *every* view class, i.e. $(\mathscr{D}A)^{\sim} = \mathscr{V}^{\sim}$? (If so, there's a *compact* example since, as can be proved constructively in a few lines, for any A there's compact C with $(\mathscr{D}A)^{\sim} \subseteq (\mathscr{D}C)^{\sim}$). Since $|\mathscr{C}X| = |\mathscr{V}| = 2^{\aleph_0}$, an answer in the negative cannot immediately be given on cardinality grounds alone. In connection with the question, preorder $\mathscr{C}X$ by $A \lesssim B \Leftrightarrow (\mathscr{D}A)^{\sim} \subseteq (\mathscr{D}B)^{\sim}$. The question then asks whether a greatest element exists. If not, consider maximal elements. Minimal elements may also be of interest; for example, X and $\emptyset$ are both minimal, having only one view class each. Note that every countable subset of $\mathscr{C}X$ has an upper bound with respect to $\lesssim$. Consider replacing $\mathscr{C}X$ in the above matters by smaller classes of sets, for example the connected closed sets, or the residual closed sets. In general, say A is __universal__ for $\mathscr{B} \subseteq \mathscr{C}X$ if A is consistent with all view classes of elements of $\mathscr{B}$; investigate such universality.

__7.4__ Note that a view space $(X,\mathscr{D})$ can be identified with the similarity view space $(X,\{Id_X\},\mathscr{D})$, the similarity group being trivial.

__Section 3__ : __View Self-Similarity__

__7.5__ Call a view **__generative__** if it is a generative view of some set (the latter concept being defined prior to 3.1). Then it's quite easily proved that a view v is generative iff it is *properly* embedded in itself, in the sense that some similarity g embeds v in v such that $\overline{g\ dom\ v} \subsetneq dom\ v$.

__7.6__ Look for examples (if there *are* any . . .) of self-similar sets which are significantly different from ω-extensions of invariant sets, examples which are not even *linked* to the matter of invariant sets. At present I know of none. Is there perhaps some iterative method for producing examples by *forcing* the self-similarity in some way? Is there a result which merely asserts the *existence* of self-similar sets with one property or another (perhaps by use of Zorn's Lemma)?

__7.7__ For self-similar A, is it necessarily true that for all $r > 0$ there's a ball of radius r which is disjoint from A (i.e. that "A contains arbitrarily large holes")?

__Section 4__ : __Self-Similarity of Some ω-Extensions of Invariant Sets__

__7.8__ The notion of ω-extensions can be used to give *explicit* examples of tilings of X by isometric copies of an invariant set K satisfying the conditions in 7.15 of Chapter 3. For suppose $\boldsymbol{i} \in F^{-\omega}$ with $\boldsymbol{i}K = X$, and for $n \geq 0$ let $T_n = \{\boldsymbol{i}{\restriction}n\,(fK) \mid f \in F^n\}$, the image under $\boldsymbol{i}{\restriction}n$ of the set of nth-level images of K, which is a tiling of $\boldsymbol{i}{\restriction}n\,K$ by isometric copies of K. Noting that in general $T_n \subseteq T_{n+1}$, $\bigcup_n T_n$ is thus a tiling of $\bigcup_n \boldsymbol{i}{\restriction}n\,K = X$ by isometric copies of K . Note also that there *exists* $\boldsymbol{i} \in F^{-\omega}$ such that $\boldsymbol{i}K = X$, by using 4.6 of the present chapter with $A = K$; for we can namely take $\boldsymbol{i} = f^{-\omega}$ for any $f \in F^{<\omega}$ with $fK \subseteq \underline{K}$ (and such f exist in abundance).

__7.9__ Concerning 4.4 and 4.6, investigate further the question of which $\boldsymbol{i} \in (hF)^{-\omega}$ give $\boldsymbol{i}D = X$.

__7.10__ In connection with 4.5, although $\boldsymbol{i}K \leftrightarrows \boldsymbol{j}K$ (and in particular, although $\boldsymbol{i}K$ is view-similar to $\boldsymbol{j}K$) $\boldsymbol{i}K$ need not be similar to $\boldsymbol{j}K$. It can be shown for example that in the case where K is the Cantor set generated by the usual two contractions,

$iK \sim jK$ iff some tail-end of i equals some tail-end of j (a tail-end of a sequence i being a sequence of the form $\langle i_n, i_{n+1}, i_{n+2}, \dots \rangle$ for some n). Thus the converse of 1.1 is false.

Section 5 : The View Topology

7.11 Using the view metric d, or more specifically its restriction to the *normal* views (see Section 6 for the definition of normal views and normalization), we can metrize $\mathscr{V}^\sim$ as follows. Define ρ on $\mathscr{V}$ by

$$\rho(v_1, v_2) = \min \{ d(f_1 v_1, f_2 v_2) \mid \text{each } f_i \text{ is a similarity mapping } dom \, v_i \text{ to } (0)_1 \}$$

$$= \min \{ d(u_1, u_2) \mid \text{each } u_i \text{ is a normal view similar to } v_i \},$$

the minimum existing by compactness of the group of rotations about 0. This is a pseudometric in which $\rho(v_1, v_2) = 0 \Leftrightarrow v_1 \sim v_2$, hence we obtain what we shall call the **view class metric** d_c on $\mathscr{V}^\sim$ (the subscript c standing for 'class') by

$$d_c(v_1^\sim, v_2^\sim) = \rho(v_1, v_2), \text{ or equivalently,}$$

$$d_c(V_1, V_2) = \min \{ d(u_1, u_2) \mid \text{each } u_i \text{ is a normal realization of } V_i \}.$$

The following proves that d_c gives the view class topology. We must show that for $U \in \mathscr{V}^*$ and $V \in \mathscr{V}$, $U \in \mu V \Leftrightarrow d_c(U, V^*) \approx 0$.

Suppose $U \in \mu V$. Taking normal $v \in V$, there's $u \in U$ with $u \in \mu v$ (as the map $\sim : \mathscr{V} \to \mathscr{V}^\sim$ is open). Where g normalizes u, then as $dom \, u \in \mu (0)_1$ it easily follows that g is near the identity map f, so by 5.14 $g u \in \mu f v = \mu v$. We thus have $d(g u, v) \approx 0$, hence as $g u \in U$ and $v \in V \subseteq V^*$, $d_c(U, V^*) \leq d(g u, v) \approx 0$.

Conversely, suppose $d_c(U, V^*) \approx 0$. Then where $d_c(U, V^*) = d(u, v)$ with $u \in U \cap (norm \, \mathscr{V})^*$ and $v \in V^* \cap (norm \, \mathscr{V})^*$, since $norm \, \mathscr{V}$ is a compact Hausdorff space with topology given by d, and $d(u, v) \approx 0$, there's $w \in norm \, \mathscr{V}$ with $u, v \in \mu w$. Thus $V^* = v^\sim \in \mu w^\sim$ giving $V = w^\sim$, and $U = u^\sim \in \mu w^\sim = \mu V$.

7.12 It is quite easily proved (using continuity of the view map) that $\mathscr{V}$ is path connected. Hence so is $\mathscr{V}^\sim$.

7.13 Call A **view class compact** if its set $(\mathscr{D} A)^\sim$ of view classes is compact (i.e. closed) in the compact Hausdorff space $\mathscr{V}^\sim$ (equivalently, if every $*$-view of A is similar to a nearstandard $*$-view of A). Which sets have this property? Note that no compact connected set does; indeed, every compact A which is view class compact has an isolated point, noting that (for any $x \in X$) the limit in $\mathscr{V}^\sim$ of $((x)_r A)^\sim$ as $r \to \infty$ is $((0)_1, \{0\})^\sim$. Any affine subspace of X is view class compact; for example a line. Less trivial, we have the following. Firstly, say $\mathscr{V} A$ is **compactly generated** if for some compact $\mathscr{E} \subseteq \mathscr{D}$, $\mathscr{V}A = (\mathscr{E}A)^{\sim}$ (every view of A is similar to a view of A with domain in $\mathscr{E}$). This implies A is view class compact (but not conversely), and may be the more significant condition of the two to study. Certain (perhaps all?) ω-extensions $\boldsymbol{i}K$ satisfying the hypothesis in 4.4 should obey this condition (stemming partly from the fact that they have generative views) and would thus be view class compact. Is *every* self-similar set view class compact? Is there any interesting special theory to be done in connection with view class compact sets? Consider also the stronger condition that the set of *oriented* view classes of A be compact in $\mathscr{V}^{\equiv}$, equivalently that the set $\{norm\ v \mid v \text{ is a view of } A\}$ of normalized views of A be compact in *norm* $\mathscr{V}$.

<u>7.14</u> Regarding the view topology on $\mathscr{V}$, if we add in the extra generating sets $[K] = \{(D,I) \mid K \subseteq \underline{I}\}$ where K is compact, we get a finer topology on $\mathscr{V}$, in which (E,J) is near (D,I) iff (E,J) is near (D,I) in the original view topology *and* $\underline{I}^{\mu} \subseteq J$. This topology is a natural one to consider in connection with views of *regular* closed sets; roughly speaking, it compares with the original view topology in the same sort of way that the body topology compares with the Vietoris topology. Further exploration is needed on its possible uses and benefits.

<u>Section 6</u> : <u>A Definition of 'Visually Fractal'</u>

<u>7.15</u> For $x,y \in A$ we say A is **<u>limit-similar at x and y</u>** if A is visually convergent at x and y to the same limit. Note that this condition is *independent* of the condition that A be similar at x and y; the latter doesn't even imply visual convergence at x and y. The equivalence classes are called the **<u>limit-similarity classes</u>** of A. Generalizing, for $a \in A$ and $b \in B$ we say A and B are **<u>limit-similar at a and b</u>** if A and B are respectively visually convergent at a and b to the same limit.

<u>7.16</u> The view metric on the *normal* views is implicitly used in the phrasing of a result of Tan Lei described in [DK] (see page 102 of Bodil Branner's contribution), concerning the visual similarity between the Mandelbrot Set M and the Julia Set J_c around a Misiurewicz point c. The result in question prompts the following definition. Suppose $a \in A \in \mathscr{C}X$ and $b \in B \in \mathscr{C}X$, and for some direct similitude h with fixed point a (i.e. h is the composition of a dilation about a and a rotation about a), $d(norm\ (a)_r\ h\ A,\ norm\ (b)_r\ B)$ converges to 0 as r decreases to 0. Then we'll say A and B are **<u>alike at a and b</u>** (which in particular holds if A and B are similar at a and b, or limit-similar at a and b). In this terminology the result of Tan Lei implies that M and J_c are alike at c and c.

<u>7.17</u> The notion of limit views can be used to give formal justification to certain informal statements about gradients in the subject of calculus. Consider for example the graph G_f in $\mathbb{R}^2$ of a continuous map $f: \mathbb{R} \to \mathbb{R}$. G_f is closed in $\mathbb{R}^2$ since f is

continuous, and we can therefore talk of views of G_f. Then where $x \in \mathbb{R}$ and $p = (x, fx)$, f is differentiable at x with derivative λ iff G_f visually converges at p to the normal view of the line through 0 with gradient λ; this then is a *formal* counterpart to saying that as we look closer and closer at p, the part of G_f we see looks more and more like the line through p with gradient λ. Note also the close connection with the idea of the 'infinitesimal microscope' described in [Ke] and attributed to K. D. Stroyan.

7.18 Limit views can be used to formulate classes of sets which may be interesting to investigate. For example, in $\mathbb{R}^2$ consider the visually convergent sets A such that for all $x \in \partial A$, $lim_A\ x$ is a view of a half-plane whose boundary line passes through 0.

7.19 As a weakening of the concept of visual convergence, we'll say that A is **weakly convergent** at $x \in A$ if $((x)_r A)^\sim$ converges in $\mathcal{V}^\sim$ as r decreases to 0. This is implied by convergence at x, but the concepts are not equivalent. For example, if A is a spiral in $\mathbb{R}^2$ with centre x, then $((x)_r A)^\sim$ is the same for *all* r, so A is weakly convergent at 0; but A is not convergent at 0.

7.20 Generalize 6.9 to $\mathbb{R}^n$.

7.21 Is it true that a view self-similar set $A \neq X$ is fractal iff it is fractal *somewhere* ?

7.22 Characterize the boundary points at which the Mandelbrot set is fractal. Conjecture: there are only countably many boundary points at which the set is *convergent*.

Miscellaneous

7.23 In matters concerned with $\mathbb{R}^n$ we concentrated on having the direct similitudes form our similarity group. Consider generalising as much of the work as possible to the case where the similarity group consists instead of all affine bijections.

7.24 The following view-related properties should be worth considering. We say A has **indeterminable scale** if every view class of A has A-realizations (realizations which are views of A) of different radius. This expresses that the scale of A cannot be determined by view classes alone. Stronger, A has **completely indeterminable scale** if every view class of A has A-realizations of arbitrarily small or large radius; for example, any ω-extension $\mathfrak{i}K$ in 4.4 has this property. For self-similar A, every view class of A at least has A-realizations of arbitrarily *small* radius. Another condition of the same ilk is that every view class of A have at least *two* A-realizations. All these conditions and others are concerned with how well a set A serves as a landmark by which one is trying to ascertain ones bearings, given view classes (views from a remote camera) alone.

<u>7.25</u> So far we have talked only of views of *sets*, but it is possible to sensibly define views of *other* types of object; for example, views of continuously coloured closed sets. For a space X, and a space Y thought of as a space of colours, we can define a <u>continuously coloured closed set</u> as a continuous map $f: A \to Y$ for some $A \in \mathscr{C}X$. Taking a view structure $\mathscr{D}$ on X, the D-view of f is then defined as $(D, f \upharpoonright D)$, $f \upharpoonright D$ being the 'object part' of the view, our 'object set' comprising the continuously coloured closed sets. The basic definitions and theory regarding views of sets largely carry over to the new case (including the matter of topologizing the view set $\mathscr{V}$ in a natural way). Note incidentally that a closed set can be thought of as a continuously coloured closed set which is constant ; we thus have a *generalization* of closed sets.

A significantly *different* area to explore and develop is that of views of Borel measures m on closed subsets A of a space X (more precisely, on the Borel algebra $\mathscr{B}A$ of A) such that the support of m is the whole of A. For such a measure, the D-view of m would be defined as $(D, m \upharpoonright D)$ where $m \upharpoonright D$ is short for $m \upharpoonright \mathscr{B}(A \cap D)$.

In view of these remarks, the question arises of whether or not we can usefully *axiomatize* the framework of 'views of objects' (along with the later additional structure of having a 'similarity group' G involved). It seems that one *can* go a significant way along these lines, but on balance I believe the matter should be placed on 'hold', pending further exploration in the various separate instances described so far (namely views of sets, coloured sets, and measures), along with any *other* significantly different instances (of which I presently know none).

Appendices

Appendix 1 : Topological Monoid Actions

Recall that a monoid consists of a set M together with an associative binary operation on M (usually just denoted by juxtaposition) and an identity element e with respect to the operation. If M is further equipped with a topology making the operation *continuous* (as a map $M \times M \rightarrow M$, giving $M \times M$ the product topology) we have a topological monoid. Suppose M is one such. Then for a topological space X, a topological monoid action of M on X is a monoid homomorphism $\theta: M \rightarrow C(X,X)$ (the latter being the monoid of continuous operations on X under composition $\circ$) whose corresponding 'evaluation map' $M \times X \rightarrow X$ is continuous. For $p \in M$, θp is the action of p, and $(\theta p)x$ is usually just written as px; the homomorphism requirement thus asks that in general $(pq)x = p(qx)$ and $ex = x$. The evaluation map sends each (p,x) to px of course. Continuity of this map (i.e. $q \in \mu p$ and $\alpha \in \mu x \Rightarrow q\alpha \in \mu\, px$) actually *implies* that the action of each $p \in M$ is continuous. Note that if X is a locally compact Hausdorff space, a monoid homomorphism $\theta: M \rightarrow C(X,X)$ is a topological monoid action iff θ is continuous, where $C(X,X)$ has the compact-open topology.

A1.1 Proposition For a Hausdorff monoid M, $\mathscr{K}M$ forms a Hausdorff monoid under the operation $PQ = \{pq \mid p \in P \text{ and } q \in Q\}$, and the map $p \mapsto \{p\}$ is a topological monoid embedding of M in $\mathscr{K}M$.

Proof:

Each PQ *is* in $\mathscr{K}M$ being the continuous image of compact $P \times Q$ under the monoid operation on M. Associativity of the new operation follows from that of the original, and the identity element is $\{e\}$. Likewise with continuity, for if $P' \in \mu P$ and $Q' \in \mu Q$ then $P' \times Q' \in \mu(P \times Q)$ in $\mathscr{K}(M \times M)$ so by continuity of the original operation $M \times M \rightarrow M$ and hence of the induced operation $\mathscr{K}(M \times M) \rightarrow \mathscr{K}M$ (see 8.1 of Chapter 1) we get $P'Q' \in \mu\, PQ$ as required. Lastly, we know the map $p \mapsto \{p\}$ embeds the space M in the space $\mathscr{K}M$, and it's also a monoid embedding since in general $\{pq\} = \{p\}\{q\}$ and $\{e\}$ = the identity element of $\mathscr{K}M$. ∘

A1.2 Proposition If a Hausdorff monoid M acts on a Hausdorff space X, then

(1) M acts on $\mathscr{K}X$ by $pA = \{pa \mid a \in A\}$.

(2) $\mathscr{K}M$ acts on $\mathscr{K}X$ by $PA = \{pa \mid p \in P \text{ and } a \in A\} = \bigcup_{p \in P} pA$.

Proof:

(2) Each PA *is* in $\mathscr{K}M$ being the continuous image of $P \times A$ under the evaluation map of the original action. $(PQ)A = \{(pq)a \mid p \in P, q \in Q, a \in A\} = \{p(qa) \mid p \in P, q \in Q, a \in A\} = P(QA)$ and $\{e\}A = \{ea \mid a \in A\} = A$ so we have a monoid action. And continuity of the evaluation map $\mathscr{K}M \times \mathscr{K}X \rightarrow \mathscr{K}X$ follows from that of the original. For if $P' \in \mu P$ and $A' \in \mu A$ then $P' \times A' \in \mu(P \times A)$ in $\mathscr{K}(M \times X)$ so by continuity

of the evaluation map $M \times X \to X$ and hence of the induced map $\mathcal{K}(M \times X) \to \mathcal{K}X$ we obtain $P'A' \in \mu PA$ as required.

(1) By (2) and the embedding of M in $\mathcal{K}M$. o

Note that if we take X to be a locally compact Hausdorff space and M to be $C(X,X)$ under composition and the compact-open topology, with M acting on X in the natural way (each $f \in M$ *being* its own action on X), then the action of $\mathcal{K}M$ on $\mathcal{K}X$ is the map $\bigcup : \mathcal{K}C(X,X) \to C(\mathcal{K}X, \mathcal{K}X)$ defined at the end of section 9 of Chapter 1, and the *continuity* of this map, noted there in 9.4, is now alternatively proved by (2) above, which gives a fuller picture of what's going on.

Appendix 2 : Finite and Infinite Sequences in a Hausdorff Space

Throughout this section let F be a Hausdorff space. A finite sequence in F is a function $n \to F$ where $n \in \omega = \{0, 1, 2 \ldots\}$; here we are treating naturals in their set-theoretic sense so n consists of all lesser m. n is the length of the above finite sequence ; the set of finite sequences in F of length n is thus F^n. In distinction to the generic symbols f and g for elements of F we may denote finite sequences in F by $\boldsymbol{f}$ and $\boldsymbol{g}$ in bold type. The length of $\boldsymbol{f}$ will be denoted by $|\boldsymbol{f}|$. The set of finite sequences in F is denoted by $F^{<\omega}$, being $\bigcup_n F^n$, and this forms a monoid under the natural operation of concatenation wherein $\boldsymbol{f}\boldsymbol{g}$ is '$\boldsymbol{f}$ followed by $\boldsymbol{g}$', a finite sequence of length $|\boldsymbol{f}| + |\boldsymbol{g}|$, the first $|\boldsymbol{f}|$ terms being those of $\boldsymbol{f}$ and the subsequent terms being those of $\boldsymbol{g}$. Algebraically this is the free monoid generated by F. The identity element is the finite sequence of length 0, namely $\emptyset$ (the 'empty sequence'). We also have a partial order $\leq$ wherein $\boldsymbol{f} \leq \boldsymbol{g} \Leftrightarrow \boldsymbol{g}$ extends $\boldsymbol{f}$, the least element being $\emptyset$. A finite sequence $\boldsymbol{f}$ of length n may sometimes be written as $\langle f_0, \ldots, f_{n-1}\rangle$ or even just $f_0 \ldots f_{n-1}$. In particular a sequence $\langle f \rangle$ of length 1 is often denoted by (or even identified with) its term f. By a sequence in F (without the term 'finite' appended or implied) we mean a function $\omega \to F$. The set of sequences in F is thus F^ω. As with finite sequences we usually use bold type $\boldsymbol{f}$ and $\boldsymbol{g}$ for sequences in F, though a sequence $\boldsymbol{f}$ may also be written as $\langle f_n \rangle$. For $\boldsymbol{f} \in F^\omega$ and $n \in \omega$, $\boldsymbol{f}\restriction n$ denotes the restriction of $\boldsymbol{f}$ to n, a finite sequence of length n.

The product topology on $F^{<\omega}$ is the unique topology on $F^{<\omega}$ in which each set of the form F^n is open and has its natural product topology. It is easily seen that . . .

A2.1 Proposition The product topology on $F^{<\omega}$ makes $F^{<\omega}$ into a Hausdorff topological monoid in which F is embedded by $f \mapsto \langle f \rangle$. ○

Under the binary operation and topology described above then, $F^{<\omega}$ can be thought of as the 'free topological monoid generated by F'.

We now consider the set $F^{\leq\omega} = F^{<\omega} \cup F^\omega$ of finite or infinite sequences in F. Extending the operation of concatenation to the case of forming the infinite sequence $\boldsymbol{f}\boldsymbol{g}$ from finite $\boldsymbol{f}$ and infinite $\boldsymbol{g}$ in the obvious way, it is clear that $F^{\leq\omega}$ forms a 'partial monoid' in the sense that concatenation is associative wherever the definition makes sense (namely, $\boldsymbol{f}(\boldsymbol{g}\boldsymbol{h}) = (\boldsymbol{f}\boldsymbol{g})\boldsymbol{h}$ whenever $\boldsymbol{f}$ and $\boldsymbol{g}$ are finite), and there is an 'identity element' having no effect under the operation (namely $\emptyset$). The *topology* on $F^{<\omega}$ can be extended to one on $F^{\leq\omega}$ in a natural way too. To facilitate the description let 0 denote some element not in F, and let $F_0 = F \cup \{0\}$ have the topology in which F forms an open subspace and $\{0\}$ is open (i.e. 0 is isolated). Now identify each $\boldsymbol{f} \in F^{<\omega}$ with $\boldsymbol{f}\,0^\omega \in F_0^{\,\omega}$ where $\boldsymbol{f}\,0^\omega$ is $\boldsymbol{f}$ followed by repeated '0's ; you can think of the '0's as indicating places in which no element of F is present. Under this correspondence $F^{\leq\omega}$ can be considered as a subspace of $F_0^{\ \omega}$, the latter naturally being given the product topology, and in this way we have described a topology on $F^{\leq\omega}$ which we'll call the <u>product topology on $F^{\leq\omega}$</u>. The salient facts are as follows, where for $f \in F^{<\omega}$ f^ω denotes the element of F^ω in which f is endlessly repeated ;

<u>**A2.2**</u> <u>**Proposition**</u> Under the product topology, $F^{\leq\omega}$ is Hausdorff and . . .

(1) Each product space F^n forms an open and closed subspace of $F^{\leq\omega}$.

(2) $F^{<\omega}$ (with the product topology) forms an open subspace of $F^{\leq\omega}$ and the product space F^ω forms a closed subspace of $F^{\leq\omega}$.

(3) For $f \in F^\omega$, $\mu f = \{g \in (F^{\leq\omega})^* \mid \omega \subseteq \text{dom } g \text{ and } \forall n \in \omega \ \ g_n \in \mu f_n\}$.

(4) For $f \in F^\omega$, $f = \lim \langle f{\restriction}n\rangle$. (Hence $F^{<\omega}$ is dense in $F^{\leq\omega}$).

(5) For $f \in F^{<\omega}$, $f^\omega = \lim \langle f^n \rangle$.

Proof:
As F_0 is Hausdorff so is $F_0^{\ \omega}$ hence so is the 'subspace' $F^{\leq\omega}$. By the formulation of the monads in $F_0^{\ \omega}$ that in general $\mu h = \{g \in (F_0^{\ \omega})^* \mid \forall m \in \omega \ \ g_m \in \mu h_m\}$, and that in F_0 $\mu 0 = \{0\}$ and (for $f \in F$) μf = the monad of f in F, it's easily seen that (3) holds, hence F^ω receives its natural product topology, and that for $f \in F^n$, $\mu f = \{g \in (F^n)^* \mid \forall m < n \ \ g_m \in \mu h_m\}$ = the monad of f in F^n, hence F^n receives its product topology and is open. The union $F^{<\omega}$ hence *also* receives its product topology and is open, the complement F^ω therefore being closed. And for $f \in F^{\leq\omega}$, if μf intersects $(F^n)^*$ then $f \in F^n$, showing that F^n is closed. (4) follows from (3) since for infinite n $f{\restriction}n \in \mu f$. And (5) is a corollary to (4), $\langle f^n \rangle$ being a subsequence of $\langle f^\omega{\restriction}n\rangle$. ○

From the description of the monads it's easily verified that the partial binary operation of concatenation is continuous wherever defined, making $F^{\leq\omega}$ in this sense a 'partial topological monoid'. Also note that the map $F^{<\omega} \rightarrow F^\omega$ sending each f to f^ω is continuous.

<u>**A2.3**</u> <u>**Proposition**</u> $F^{\leq\omega}$ is compact $\Leftrightarrow$ F is compact.

Proof:
$\Rightarrow$: $F \cong F^1$, and the latter is closed in $F^{\leq\omega}$, hence compact.
$\Leftarrow$: Then F_0 is compact hence so is $F_0^{\ \omega}$. And it's easily seen that $F^{\leq\omega}$ is *closed* in $F_0^{\ \omega}$, hence is compact too. ○

We could alternatively have proved the above by using . . .

A2.4 **Note** For $g \in (F^{\leq\omega})^*$,

g is nearstandard $\Leftrightarrow$ for all finite $n \in \operatorname{dom} g$, g_n is nearstandard.

Proof:

$\Rightarrow$: Let $g \in \mu f$. If $f \in F^{\omega}$ then $\forall n \in \omega$, $g_n \in \mu f_n$ so g_n is nearstandard. If $f \in F^m$ then $\operatorname{dom} g = m$ and $\forall n < m$, $g_n \in \mu f_n$ so g_n is nearstandard.

$\Leftarrow$: Suppose $\omega \subseteq \operatorname{dom} g$. Then letting $f \in F^{\omega}$ with $\forall n \in \omega \;\; g_n \in \mu f_n$, we have $g \in \mu f$ so g is nearstandard. On the other hand suppose $\operatorname{dom} g = m \in \omega$. Then letting $f \in F^m$ with $\forall n < \omega \;\; g_n \in \mu f_n$, we have $g \in \mu f$ so g is again nearstandard. o

Lastly we show that for compact F, analogous to the fact that each infinite sequence is the limit of its finite restrictions, F^{ω} and $F^{\leq\omega}$ are each the limit (with respect to the Vietoris topology on $\mathscr{K}\, F^{\leq\omega}$) of their 'finite restrictions' ;

A2.5 **Proposition** For compact F, with respect to the Vietoris topology on $\mathscr{K}\, F^{\leq\omega}$

(1) $F^{\omega} = \lim \langle F^n \rangle$.

(2) $F^{\leq\omega} = \lim \langle F^{\leq n} \rangle$ where $F^{\leq n} = \bigcup \{F^m \mid m < n\}$.

Proof:

First note that as F is compact so are F^{ω}, F^n, $F^{\leq\omega}$, and $F^{\leq n}$.

(1) We must show that for infinite n, where S_n is the set of internal sequences in F^* of length n, $S_n \in \mu\, F^{\omega}$. By compactness of F, every element of S_n is near an element of F^{ω}. Conversely, for all $f \in F^{\omega}$, $f^* \restriction n$ belongs to S_n and is near f. So $S_n \in \mu\, F^{\omega}$ as required.

(2) Similar sort of proof to the above. o

Appendix 3 : **Continuity of fix : *Contrac X* →*X***

In this section we consider the possible continuity of the fixed point map fix : *Contrac X* →*X* with respect to the compact-uniform and bounded-uniform topologies on the set *Contrac X* of contractions of a nonempty complete metric space *X*. If *X* is locally compact the compact-uniform topology is sufficient for continuity to hold, and this can be proved by purely topological means as follows.

For a Hausdorff space X and $f \in C(X,X)$, we say x is a **topologically attractive** fixed point of f if for all $y \in X$, $\langle f^n y\rangle \to x$, which indeed *implies* x is fixed under f as f is continuous. Such a point will be the *unique* fixed point of f, and denoted by **fix** f.

A3.1 Proposition Let X be a locally compact Hausdorff space and $F = \{f \in C(X,X) \mid f$ has a topologically attractive fixed point$\}$. Then the natural fixed point map fix : $F \to X$ is continuous with respect to the compact-open topology on F.

Proof:
Let $f \in F$ and $g \in F^*$ with $g \in \mu f$. Let $x = \text{fix}\, f$ and suppose for a contradiction that fix $g \notin \mu x$. Then let C be a compact neighbourhood of x with fix $g \notin C^*$. Now $\neg C^*$ is a $*$-neighbourhood of fix g, so as $\langle g^n x\rangle_{n \in \omega^*}$ converges to fix g in X^*, there's $n \in \omega^*$ such that $\forall m \geq n \;\; g^n x \notin C^*$. Let n be the *least* such. Note that $n > 1$ as $g x \in \mu f x = \mu x \subseteq C^*$. By leastness of n, $g^{n-1}x \in C^*$ so by compactness of C let $c \in C$ with $g^{n-1}x \in \mu c$. As $\langle f^m c\rangle \to x$ let $m \geq 1$ with $f^m c \in \underline{C}$. By continuity of composition in $C(X,X)$, $g^m \in \mu f^m$ so $g^m(g^{n-1}x) \in \mu f^m c$, i.e. $g^{n+m-1}x \in \mu f^m c$ giving (as $\mu f^m c \subseteq C^*$) $g^{n+m-1}x \in C^*$, a contradiction as $n+m-1 \geq n$. ○

Assume from now on that X is a nonempty complete metric space. Since the compact-uniform topology on $C(X,X)$ is the compact-open topology, and since in the above result *Contrac* $X \subseteq F$, it follows that . . .

A3.2 Corollary If X is locally compact, fix : *Contrac* $X \to X$ is continuous with respect to the compact-uniform topology. ○

In Theorem 2 of [Nad], *sequential* continuity of fix : *Contrac* $X \to X$ is proved, namely that if $\langle f_n\rangle$ converges to g then $\langle \text{fix}\, f_n\rangle$ converges to fix g. For *countably* based X this is *equivalent* to continuity since *Contrac* X (moreover $C(X,X)$) is then *also* countably based (see [Du], page 265, 5.2). In his result Nadler uses the topology of pointwise convergence, but on *Contrac* X this coincides with the compact-uniform topology, as pointed out in 'Preliminaries'.

Without local compactness, continuity with respect to the compact-uniform topology may fail, one example being given in [Li,1] along with another proof of A3.2.

However, by placing an upper bound u on the Lipschitz ratios allowed, we do at least have a restricted version. For $u \in [0,1)$ let $\mathit{Contrac}_u X = \{f \in \mathit{Contrac}\, X \mid r_f \leq u\}$.

<u>**A3.3**</u> <u>**Lemma**</u> For $g \in \mathit{Contrac}\, X^*$ with $r_g \not\approx 1$, $\forall x \in X$, $g\,x \approx x \Leftrightarrow x \approx \mathrm{fix}\, g$.

Proof:

$\Rightarrow$: Then by transfer of a standard inequality, $d(x, \mathrm{fix}\, g) \leq d(x, g\,x)/(1-r_g) \approx 0$ as $1-r_g$ is finite and noninfinitesimal and $d(x,g\,x) \approx 0$.

$\Leftarrow$: Then $g\,x \approx \mathrm{fix}\, g$ too, so $g\,x \approx x$. ○

<u>**A3.4**</u> <u>**Proposition**</u> For any $u < 1$, $\mathrm{fix} : \mathit{Contrac}_u X \to X$ is continuous with respect to the compact-uniform topology.

Proof:

Let $f \in \mathit{Contrac}_u X$ and $g \in \mathit{Contrac}_u X^*$ with $g \in \mu f$. Then where $x = \mathrm{fix}\, f$, $g\,x \approx f\,x = x$ so by the last lemma $x \approx \mathrm{fix}\, g$ as required. ○

The above is also proved in [Li,1]. As an alternative to reducing the domain of the fixed point map in order to obtain continuity, we could instead weaken the topology on *Contrac X*. Weakening to the bounded-uniform topology proves to be sufficient, which follows from the next result along with another proof of A3.2. Recall from Section 6 of Chapter 1 that an ideal $\mathcal{J}$ on X is said to be 'topological' if every element of X has a neighbourhood belonging to $\mathcal{J}$.

<u>**A3.5**</u> <u>**Proposition**</u> For any topological ideal $\mathcal{J}$ on X, $\mathrm{fix} : \mathit{Contrac}\, X \to X$ is continuous with respect to the $\mathcal{J}$-uniform topology on *Contrac X*.

Proof:

Let $f \in \mathit{Contrac}\, X$ and $g \in \mathit{Contrac}\, X^*$ with $g \in \mu f$. Letting $x = \mathrm{fix}\, f$, since x has a neighbourhood N belonging to $\mathcal{J}$ then for all sufficiently small $\epsilon > 0$, letting $A = [x]_\epsilon$ we have $A \subseteq N$ so $A^* \subseteq \nu\mathcal{J}$, and we find that A^* is closed under g;

$$
\begin{aligned}
g\,A^* &\subseteq (f^* A^*)^{\approx} && \text{as } \forall a \in A^*\ g\,a \approx f^* a \\
&= ((f A)^*)^{\approx} \\
&\subseteq ([x]_{r_f \epsilon}{}^*)^{\approx} && \text{as } f A \subseteq [x]_{r_f \epsilon} \\
&\subseteq A^* && \text{as } r_f \epsilon < \epsilon .
\end{aligned}
$$

So as A^* is also $*$-closed, transfer of a standard fact gives $\mathrm{fix}\, g \in A^*$, hence $d(\mathrm{fix}\, g, x) \leq \epsilon$. So $\mathrm{fix}\, g \approx x$ as required. ○

If X is locally compact, *subCp X* is topological so the above gives an alternative proof of A3.2. Meanwhile the ideal *Bd X* of *bounded* sets is *always* topological, so . . .

<u>**A3.6**</u> <u>**Corollary**</u> $\mathrm{fix} : \mathit{Contrac}\, X \to X$ is continuous with respect to the bounded-uniform topology. ○

Appendix 4 : Reductions of a Metric Space

Throughout let X be a nonempty metric space. For the next paragraph Y denotes a metric space too, though we're mainly interested in the case $Y = X$.

A control of $f: X \to Y$ is an increasing operation δ on $[0,\infty)$ such that $\forall x,y \in X$ $d(fx,fy) \leq \delta\, d(x,y)$, equivalently $\forall \epsilon \geq 0\ \forall x,y \in X\ (\ d(x,y) \leq \epsilon \Rightarrow d(fx,fy) \leq \delta\epsilon\ )$. The idea is that δ gives a distance-wise upper bound on the expansiveness of f. For example, for a map f with finite Lipschitz ratio r_f, multiplication by r_f is a control of f. If f has a control we'll say f is controlled, the nonstandard formulation being that f^* is 'macrocontinuous', i.e. that $\forall x,y \in X^*$, if x is finitely distant from y then fx is finitely distant from fy. If f is controlled it has a *least* control, as follows. For $f: X \to Y$ define $\delta_f : [0,\infty) \to [0,\infty]$ by $\delta_f\, \epsilon = \bigvee \{\ d(fx,fy) \mid d(x,y) \leq \epsilon\ \}$. This is clearly the least increasing function $\delta : [0,\infty) \to [0,\infty]$ such that $\forall \epsilon \geq 0\ \forall x,y \in X$ $(\ d(x,y) \leq \epsilon \Rightarrow d(fx,fy) \leq \delta\epsilon\ )$, i.e. $\forall x,y \in X\ \ d(fx,fy) \leq \delta\, d(x,y)$, and it follows that f is controlled iff δ_f is finite-valued, in which case δ_f is the *least* control of f. δ_f is known in the literature as the 'modulus of continuity' of f. Note incidentally that f is controlled iff $\forall \epsilon > 0\ \exists \delta > 0\ \forall x,y \in X\ (\ d(x,y) \leq \epsilon \Rightarrow d(fx,fy) \leq \delta\ )$; compare with the definition of uniform continuity (and compare the nonstandard formulations, macrocontinuity and microcontinuity).

A technical result we'll be using later is the following, where an operation δ on $[0,\infty)$ is said to be *lower continuous* (also known as *right continuous*) if whenever $\langle a_n \rangle$ decreases to b then $\langle \delta_f a_n \rangle$ converges to $\delta_f b$.

A4.1 Note For continuous $f: X \to Y$ with X compact,

(1) $\forall \epsilon \geq 0,\ \delta_f \epsilon = \max \{ d(fx,fy) \mid d(x,y) \leq \epsilon \}$.

(2) δ_f is lower continuous.

Proof:

(1) By a nonstandard result on l.u.b.s in $[0,\infty]$, $\delta_f \epsilon = {}^\circ d(fx,fy)$ for some $x,y \in X^*$ with $d(x,y) \leq \epsilon$. But then $d({}^\circ x, {}^\circ y) = {}^\circ d(x,y) \leq \epsilon$ and $d(f\ {}^\circ x, f\ {}^\circ y) = d({}^\circ fx, {}^\circ fy) = {}^\circ d(fx,fy) = \delta_f \epsilon$.

(2) Let $\epsilon \in [0,\infty)$ and $\gamma \approx \epsilon$ with $\gamma \geq \epsilon$; we must show $\delta_f \gamma \approx \delta_f \epsilon$, i.e. (as $\delta_f \epsilon \leq \delta_f \gamma$) ${}^\circ \delta_f \gamma \leq \delta_f \epsilon$. By transfer of (1) let $x,y \in X^*$ with $d(x,y) \leq \gamma$ and $\delta_f \gamma = d(fx,fy)$. Then $d({}^\circ x, {}^\circ y) = {}^\circ d(x,y) \leq {}^\circ \gamma = \epsilon$ so $d(f\ {}^\circ x, f\ {}^\circ y) \leq \delta_f \epsilon$, i.e. $d({}^\circ fx, {}^\circ fy) \leq \delta_f \epsilon$, i.e. ${}^\circ d(fx,fy) \leq \delta_f \epsilon$, i.e. ${}^\circ \delta_f \gamma \leq \delta_f \epsilon$ as required. ○

Note that for any control δ of $f: X \to X$, for nonempty bounded $A \subseteq X$ we have $\operatorname{diam} fA \leq \delta \operatorname{diam} A$, giving by induction that $\operatorname{diam} f^n A \leq \delta^n \operatorname{diam} A$ for all n. Hence if $\langle \delta^n \operatorname{diam} A \rangle$ converged to 0, so would $\langle \operatorname{diam} f^n A \rangle$. An interest in this condition leads us to consider the following definition.

We'll say an operation δ on $[0,\infty)$ has **attractor 0** if $\forall\epsilon \geq 0 \ \langle\delta^n\epsilon\rangle \to 0$. If δ is also increasing (i.e. in general $\epsilon \leq \gamma \Rightarrow \delta\epsilon \leq \delta\gamma$) note that $\delta 0 = 0$, and the formulations below hold.

A4.2 Proposition For an increasing operation δ on $[0,\infty)$,

(1) δ has attractor 0 $\Leftrightarrow \forall\epsilon > 0$ ($\delta\epsilon < \epsilon$ and $\exists\gamma > \epsilon \ \ \delta\gamma \leq \epsilon$)

(2) If δ is lower continuous, δ has attractor 0 $\Leftrightarrow \forall\epsilon > 0 \ \ \delta\epsilon < \epsilon$.

Proof:

(2) is a simple corollary to (1), which is proved as follows.

$\Rightarrow$: Let $\epsilon > 0$. If $\delta\epsilon \geq \epsilon$ then as δ is increasing, by induction we'd have $\forall n \ \ \delta^n\epsilon \geq \epsilon$, contradicting $\langle\delta^n\epsilon\rangle \to 0$. So $\epsilon < \delta\epsilon$. Secondly, if $\forall\gamma > \epsilon \ \ \delta\gamma > \epsilon$, then by induction, for $\gamma > \epsilon$ we'd have $\forall n \ \ \delta^n\gamma > \epsilon$ contradicting $\langle\delta^n\gamma\rangle \to 0$. So $\exists\gamma > \epsilon \ \ \delta\gamma \leq \epsilon$.

$\Leftarrow$: First note that $\delta 0 = 0$, since $\forall\epsilon > 0 \ \ \delta 0 \leq \delta\epsilon < \epsilon$ (using that δ is increasing). It remains to show that for $\epsilon > 0$, $\langle\delta^n\epsilon\rangle \to 0$. Suppose $\langle\delta^n\epsilon\rangle$ does *not* converge to 0. Then no $\delta^n\epsilon$ can be 0, so $\langle\delta^n\epsilon\rangle$ is strictly decreasing since $\forall r > 0 \ \ \delta r < r$. So let $r = \lim \langle\delta^n\epsilon\rangle = \bigwedge_n \delta^n\epsilon$. As $r > 0$ there's $\gamma > r$ with $\delta\gamma \leq r$, and since for some n $\delta^n\epsilon \leq \gamma$ then also $\delta^{n+1}\epsilon = \delta\,\delta^n\epsilon \leq \delta\gamma \leq r$ giving $\delta^{n+1}\epsilon \leq r$, a contradiction. So $\langle\delta^n\epsilon\rangle \to 0$ as required. ○

We'll say $f: X \to X$ is a **reduction** of X if it has a control with attractor 0, in which case δ_f is the *least* such control (being sandwiched as it is underneath *all* controls of f). By the last result note that . . .

A4.3 Corollary For $f: X \to X$,

$$f \text{ is a reduction} \Leftrightarrow \forall\epsilon > 0 \ (\ \delta_f\epsilon < \epsilon \text{ and } \exists\gamma > \epsilon \ \ \delta_f\gamma \leq \epsilon \ )$$

Proof:

$\Rightarrow$: By the last result, putting $\delta = \delta_f$.

$\Leftarrow$: Then δ_f is finite-valued, hence is a control of f, and by the last result δ_f has attractor 0. ○

A4.4 Note Any composition of reductions is a reduction.

Proof:

If f and g are reductions, so is $f\circ g$ since $\delta_{f\circ g} \leq \delta_f \circ \delta_g \leq \delta_g$. ○

The next result compares the property of being a reduction with various other contractivity conditions, for which some definitions are in order. Encountered in [Ha] is the definition that $f: X \to X$ is a **weak contraction** if $\forall\epsilon > 0 \ \exists\gamma > \epsilon \ \ \delta_f\gamma < \epsilon$. Hata notes that every contraction is a weak contraction but not conversely. $f: X \to X$ is said (in the terminology of some of the literature of fixed point theory) to be **contractive** if for all distinct $x,y \in X$, $d(fx,fy) < d(x,y)$.

A4.5 Proposition Below, for $f\colon X \to X$ the implications $(1) \Rightarrow (2) \Rightarrow (3) \Rightarrow (4)$ hold, with equivalence if X is compact.

(1) f is a weak contraction.

(2) f is a reduction.

(3) $\forall \epsilon > 0 \quad \delta_f \epsilon < \epsilon$.

(4) f is contractive.

Proof:

$(1) \Rightarrow (2)$: Then also, $\forall \epsilon > 0 \quad \delta_f \epsilon < \epsilon$ (for where $\gamma > \epsilon$ with $\delta_f \gamma < \epsilon$, we have $\delta_f \epsilon \le \delta_f \gamma < \epsilon$. So by A4.3, f is a reduction.

$(2) \Rightarrow (3)$: By A4.3.

$(3) \Rightarrow (4)$: Then for all distinct $x,y \in X$, $d(fx,fy) \le \delta_f d(x,y) < d(x,y)$.

Assuming now that X is compact and (4) holds, we prove (1). Let $\epsilon > 0$. By contractivity of f and A4.1(1), we have $\delta_f \epsilon < \epsilon$. And since by A4.1(2) δ_f is lower-continuous, it follows that $\exists \gamma > \epsilon \quad \delta_f \gamma < \epsilon$ as required. ○

As a corollary, note that reductions are continuous since contractive maps are. The following example shows that reductions need *not* be weak contractions. Let $X = \{x_n \mid n \ge 1\} \cup \{y_n \mid n \ge 1\} \cup \{p\}$ where the symbols involved denote distinct points. Let X have the post-office metric about p in which $d(x_n, p) = 1 + \epsilon_n$ and $d(y_n, p) = 1$, where $\epsilon_n = 1/2^n$. Being a 'post-office metric about p' means that for $a \ne b$, $d(a,b) = d(a,p) + d(p,b)$. Note that X is bounded and also complete since distinct points are distance ≥ 1 apart. This fact also implies that every point is isolated, so in particular X is locally compact. Define $f\colon X \to X$ by $fx_n = y_n$, $fy_n = p$, and $fp = p$. By checking the various cases carefully it can be shown that δ_f is 0 on $[0,1]$, 1 on $(1,2]$, and 2 on $(2,\infty)$. Thus for $\epsilon > 2$, $\delta_f^{\,3}\epsilon = \delta_f^{\,2}\, 2 = \delta_f 1 = 0$, so $\langle \delta_f^{\,n} \epsilon \rangle$ is eventually 0 hence converges to 0. So f is a reduction. However, f is not a weak contraction since there is no $\gamma > 2$ with $\delta_f \gamma < 2$.

We now show that Banach's contraction mapping theorem generalizes to reductions. Note that we say a sequence $\langle a_n \rangle$ in $[0,\infty)$ *strictly* decreases to 0 if it's a decreasing sequence converging to 0 and for all nonzero a_n, $a_{n+1} < a_n$.

A4.6 Lemma For a reduction f of X,

(1) For all nonempty bounded $A \subseteq X$, $\langle \operatorname{diam} f^n A \rangle$ strictly decreases to 0.

(2) For all $x,y \in X$, $\langle d(f^n x, f^n y) \rangle$ strictly decreases to 0.

Proof:

(1) In general $\operatorname{diam} f^n A \le \delta_f^{\,n} \operatorname{diam} A$, so $\langle \operatorname{diam} f^n A \rangle \to 0$ as $\langle \delta_f^{\,n} \operatorname{diam} A \rangle \to 0$. As f is contractive, $\langle \operatorname{diam} f^n A \rangle$ is decreasing; and for any nonzero term $\operatorname{diam} f^n A$, $\operatorname{diam} f^{n+1} A = \operatorname{diam} f(f^n A) \le \delta_f \operatorname{diam} f^n A < \operatorname{diam} f^n A$.

(2) By (1) with $A = \{x,y\}$, noting $\operatorname{diam} f^n A = \operatorname{diam} \{f^n x, f^n y\} = d(f^n x, f^n y)$. ○

A4.7 **Proposition** For complete X, every reduction f of X has a metrically attractive fixed point, i.e. a fixed point p (necessarily unique) such that for all $x \in X$ $\langle d(f^n x, p)\rangle$ strictly decreases to 0.

Proof:

Let $x \in X$. Letting $x_n = f^n x$, we first show that $\langle x_n \rangle$ is a Cauchy sequence.

Putting $y = fx$ in (2) of the last lemma, $\langle d(x_n, x_{n+1})\rangle$ decreases to 0, so for all infinite m we have $x_m \approx x_{m+1}$. Suppose $\langle x_n \rangle$ is *not* Cauchy. Then let $\epsilon > 0$ such that $\forall m\ \exists n > m\ d(x_m, x_n) > \epsilon$. Taking infinite $m \in \omega^*$ and letting n be the *least* $n > m$ with $d(x_m, x_n) > \epsilon$, then $d(x_m, x_{n-1}) \leq \epsilon$ so $d(fx_m, fx_{n-1}) \leq \delta_f \epsilon$, so $\delta_f \epsilon \geq {}^\circ d(fx_m, fx_{n-1}) = {}^\circ d(x_{m+1}, x_n) = {}^\circ d(x_m, x_n) \geq \epsilon$ (using along the way that $x_m \approx x_{m+1}$) contradicting that $\delta_f \epsilon < \epsilon$. So, $\langle x_n \rangle$ is Cauchy.

So by completeness of X, $\langle x_n \rangle$ converges, say to p, which by continuity of f is therefore a fixed point of f. And for all $y \in X$, $\langle d(f^n y, p)\rangle = \langle d(f^n y, f^n p)\rangle$ which by (2) of the last lemma strictly decreases to 0, showing that p is metrically attractive. o

Note that for compact X the above result says that every *contractive* map on X has a metrically attractive fixed point ; this result appears in [Ed].

It is worth pointing out a dynamic difference between contractions and reductions. For $f: X \to X$ we define the **length** of the path $\langle f^n x \rangle$ of x under f to be $\sum_{n \geq 0} d(f^n x, f^{n+1} x)$, which if f is a contraction, is *finite* (*hence* $\langle f^n x \rangle$ is Cauchy, which then leads to Banach's contraction mapping theorem). However, under a reduction, or even a 'weak contraction', paths may have *infinite* length, so that intuitively, although as a point traces out its path it converges to a limit (assuming X is complete), it may travel an infinite distance along the way. Here is an example. Let $X = \{x_n \mid n \geq 1\} \cup \{p\}$ where the symbols involved denote distinct points, and give X the post-office metric about p in which $d(x_n, p) = 1/n$. Define $f: X \to X$ by $fp = p$ and $fx_n = x_{n+1}$. Then f is contractive hence (as X is compact) a weak contraction, but the path $\langle x_n \rangle_{n \geq 1}$ of x_1 has infinite length.

For the remainder of the section assume X is complete and let ***Reduc X*** denote the set of reductions of X. We now consider the possible continuity of the fixed point map **fix** : *Reduc* $X \to X$. Basically all we have to do is modify some of the proofs of the corresponding results for contractions given in Appendix 3. For a start, with the same proof as that of A3.2,

A4.8 **Proposition** If X is locally compact, fix : *Reduc* $X \to X$ is continuous with respect to the compact-uniform topology. o

The next two results parallel A3.3 and A3.4. For any increasing operation u on $[0,\infty)$ with attractor 0, let $\boldsymbol{Reduc}_u X = \{f \in Reduc\ X \mid u$ is a control of f, i.e. $\delta_f \le u\}$.

__A4.9__ __Lemma__ For any $g \in Reduc\ X^*$ such that for all real $\epsilon > 0$ $\delta_g\, \epsilon \not\approx \epsilon$,

$$\forall x \in X,\quad g\,x \approx x \Leftrightarrow x \approx \text{fix}\ g\,.$$

Proof:

$\Rightarrow$: Suppose $x \not\approx \text{fix}\ g$; let $\epsilon > 0$ be real with $\text{fix}\ g \notin [x]_\epsilon{}^*$. In X^* the $*$-sequence $\langle g^n x\rangle_{n \in \omega^*}$ converges to fix g, so for all sufficiently large $n \in \omega^*$, $g^n x \notin [x]_\epsilon{}^*$. Let n be the *least* such. $n > 1$ since $g\,x \approx x$. By leastness of n, $d(g^{n-1}x,x) \le \epsilon$ so $d(g^n x, g\,x) \le \delta_g\,\epsilon$ so $^\circ d(g^n x,g\,x) \le {}^\circ\delta_g\,\epsilon < \epsilon$. But as $g\,x \approx x$ then $^\circ d(g^n x,x) = {}^\circ d(g^n x,g\,x) < \epsilon$ which contradicts $d(g^n x,x) > \epsilon$. So $x \approx \text{fix}\ g$ after all.

$\Leftarrow$: Then $g\,x \approx \text{fix}\ g$ too, so $g\,x \approx x$. ○

__A4.10__ __Proposition__ For any increasing operation u on $[0,\infty)$ with attractor 0, fix : $Reduc_u X \to X$ is continuous with respect to the compact-uniform topology.

Proof:

Let $f \in Reduc_u X$ and $g \in Reduc_u X^*$ with $g \in \mu f$. Then where $x = \text{fix}\ f$, $g\,x \approx f\,x = x$ so by the last lemma (noting that for all real $\epsilon > 0$, $\delta_g\,\epsilon \le u\epsilon < \epsilon$ giving $\delta_g\,\epsilon \not\approx \epsilon$), $x \approx \text{fix}\ g$ as required. ○

Finally, in the same way as in Appendix 3, the following provides an alternative proof of continuity with respect to the compact-uniform topology, and gives continuity with respect to the bounded-uniform topology ;

__A4.11__ __Proposition__ For any topological ideal $\mathcal{J}$ on X, fix : $Reduc\ X \to X$ is continuous with respect to the $\mathcal{J}$-uniform topology on $Reduc\ X$.

Proof:

Identical to the proof of A3.5, except replacing $r_f\epsilon$ by $\delta_f\epsilon$. ○

__A4.12__ __Corollary__ fix : $Reduc\ X \to X$ is continuous with respect to the bounded-uniform topology. ○

Appendix 5 : Nonoverlapping Sets, and Tilings

Until further notice let X be a Hausdorff space. Recall that the *boundary* of a closed set A is $\partial A = A - \underline{A}$. We'll say closed sets A and B are **nonoverlapping** in X if they can intersect only at their boundaries, i.e. if $A \cap B \subseteq \partial A \cap \partial B$, *equivalently* if $A \cap \underline{B} = \emptyset$ and $B \cap \underline{A} = \emptyset$. This implies that $\underline{A}$ and $\underline{B}$ are disjoint, whilst being *equivalent* to the latter if A and B are regular. Note also that if A and B are nonoverlapping, so are A' and B' for any closed $A' \subseteq A$ and $B' \subseteq B$. If U and V are disjoint open sets, $\overline{U}$ and $\overline{V}$ are nonoverlapping. *Residual* closed sets are trivially nonoverlapping. We'll say a family $\langle A_i \mid i \in I \rangle$ of closed subsets of X is **nonoverlapping** in X if $i \neq j \Rightarrow A_i$ and A_j are nonoverlapping.

A5.1 Proposition For $A,B \subseteq C \subseteq X$ with A and B closed in X,

A and B are nonoverlapping in $C \Rightarrow A$ and B are nonoverlapping in X,

with the converse holding if A and B are regular.

Proof:

$\Rightarrow$: Where $\underline{\ \ }_C$ denotes 'interior in C' we have $\underline{A} \subseteq \underline{A}_C$, so where ∂_C denotes 'boundary in C', $\partial_C A \subseteq \partial A$. Likewise $\partial_C B \subseteq \partial B$. Hence $A \cap B \subseteq \partial_C A \cap \partial_C B \subseteq \partial A \cap \partial B$.

$\Leftarrow$: A is disjoint from $\underline{B}_C$ since if it *intersected* $\underline{B}_C$ (which is open in C) so would $\underline{A}$ (as $\underline{A}$ is dense in A) hence $\underline{A}$ would intersect B, a contradiction. Likewise B is disjoint from $\underline{A}_C$, so A and B are nonoverlapping in C. o

We say a family $\langle A_i \mid i \in I \rangle$ of subsets of X is a **tiling** of X if it's a nonoverlapping cover of X by nonempty regular closed sets. This can be reformulated as the property (2) below of 'strong nonoverlapping' which *automatically* implies the regularity condition (that each A_i be regular) ;

A5.2 Note For a cover $\langle A_i \mid i \in I \rangle$ of X by nonempty closed sets, the following are equivalent ;

(1) $\langle A_i \mid i \in I \rangle$ is a tiling of X.

(2) $\forall i \in I$ A_i and $\overline{\bigcup_{j \neq i} A_j}$ are nonoverlapping.

(3) $\forall i \in I$ A_i is disjoint from the interior of $\overline{\bigcup_{j \neq i} A_j}$.

Proof:

(1) $\Rightarrow$ (3) : Let $i \in I$ and $B = \overline{\bigcup_{j \neq i} A_j}$. If A_i *intersected* the interior of B then so would the interior V of A_i (being dense in A_i as A_i is regular), thus B would intersect V, hence so would $\bigcup_{j \neq i} A_j$ (as this is dense in B), hence some A_j $(j \neq i)$ would intersect V, contradicting that A_j and A_i do not overlap.

(3) $\Rightarrow$ (2) : Take i and B as above. We need to show that B is disjoint from the interior V of A_i. If B *intersected* V, so would $\bigcup_{j \neq i} A_j$, hence so would some A_j $(j \neq i)$, hence

A_j would intersect the interior of $\overline{\bigcup_{k \neq j} A_k}$ (as this expands V), contradicting (3).

(2) ⇒ (1) : Since in general $A_i \subseteq \overline{\bigcup_{j \neq i} A_j}$, we see that $\langle A_i \mid i \in I \rangle$ is nonoverlapping, so it remains to show that each A_i is regular. Let $B = \overline{\bigcup_{j \neq i} A_j}$ and V be the interior of A_i. The complement $\neg B$ of B is an open subset of A_i so $\neg B \subseteq V$; and as A_i is disjoint from $\underline{B}$ we have $A_i \subseteq \neg \underline{B} = \overline{\neg B} \subseteq \overline{V}$, so A_i is regular as required. ○

A5.3 Corollary If A_1 and A_2 are nonoverlapping nonempty closed sets with union X, then A_1 and A_2 are regular.

Proof:

$\langle A_i \mid i \in I \rangle$ obeys (2) above, hence obeys (1), so each A_i is regular. ○

So far we've been talking of tilings as families, but for some purposes it is more natural to consider tilings in the form of sets. For the remainder of this appendix a **tiling** of X will mean a *set* $\mathscr{A}$ of mutually nonoverlapping regular closed sets which covers X. (Of course, $\mathscr{A} \subseteq \mathscr{P}X$ is a tiling of X iff its corresponding family $\langle A \mid A \in \mathscr{A} \rangle$ is a tiling of X in the earlier definition.) The following, used in 7.1 of Chapter 3, says that if $\mathscr{A}$ is compact in the Vietoris topology on $\mathscr{C}X$ then $\mathscr{A}$ must be finite ;

A5.4 Proposition With respect to the Vietoris topology on $\mathscr{C}X$, every compact tiling of X is finite.

Proof:

Suppose $\mathscr{A}$ is infinite. Then let B be a nonstandard element of $\mathscr{A}^*$, and by compactness of $\mathscr{A}$ let $A \in \mathscr{A}$ with $B \in \mu A$. Letting $C = \bigcup(\mathscr{A} - \{A\})$, then since $B \in \mathscr{A}^* - \{A^*\}$ we have $B \subseteq \bigcup(\mathscr{A}^* - \{A^*\}) = \bigcup((\mathscr{A} - \{A\})^*) = C^*$, hence $A \subseteq st\, B \subseteq st\, C^* = \overline{C}$, contradicting that A and $\overline{C}$ are nonoverlapping (essentially given by (2) of A5.2). ○

We conclude this appendix with a nonstandard proof of an existence theorem for tilings, bringing various hyperspace work into play. Throughout let $X = \mathbb{R}^n$, $\mathfrak{B}X$ be the set of bodies in X under the body topology (see Appendix 6), and $\mathscr{T} \in \mathscr{K}\mathfrak{B}X$. We shall be concerned with the existence of a tiling of X by copies of elements of $\mathscr{T}$, 'copy' being used in the following variable sense. Let G be any group of isometries of X which is closed in the group *Sim* X of similitudes of X under the compact-open topology; for example G could consist of *all* isometries, or just the *direct* isometries, or just the translations. Then letting $\sim$ be the equivalence relation in $\mathfrak{B}X$ induced by the action of G on $\mathfrak{B}X$, we'll say B is a **copy** of A if $B \sim A$.

Roughly speaking, the tiling existence theorem states that if we can do an arbitrarily good job of tiling over *bounded* subsets of X with copies of elements of $\mathscr{T}$, there exists a tiling of X by copies of elements of $\mathcal{S}$. To state this more precisely we have our last few definitions. A **packing** of X is a set of mutually nonoverlapping nonempty regular closed subsets of X; thus a *tiling* of X is just a packing of X which covers X. If the sets involved are all copies of elements of $\mathcal{S}$ we shall respectively refer to a **$\mathcal{S}$-packing** and **$\mathcal{S}$-tiling** of X. As a weakening of the notion of covering, we'll say a $\mathcal{S}$-packing $\mathcal{A}$ of X covers $B \subseteq X$ **to within tolerance ϵ** if $B \subseteq [\bigcup \mathcal{A}]_\epsilon$. Then our tiling existence theorem states that if every bounded subset of X can be covered by $\mathcal{S}$-packings to within arbitrarily small tolerances, there exists a $\mathcal{S}$-tiling of X. In particular we'll have the trivial corollary that if every bounded subset of X can be covered by a $\mathcal{S}$-packing then there exists a $\mathcal{S}$-tiling of X; a restricted version of this is given in [GS] for $X = \mathbb{R}^2$ and $\mathcal{S}$ being a finite set of topological discs.

The basic idea of the nonstandard proof is similar to that used for example in the nonstandard proof of the graph-colouring theorem that if every finite subgraph of a graph G is n-colourable, the *whole* graph G is n-colourable. There, one takes a hyperfinite intermediary H of G, for which by transfer there's an n-colouring, and one then simply restricts this colouring to G to obtain an n-colouring of G. The situation for tilings is more involved but has the same flavour. Assuming every bounded subset of X can be covered to arbitrary tolerances by $\mathcal{S}$-packings, then taking a $*$-bounded set B expanding $bd\, X^*$, by transfer there's a $*$-($\mathcal{S}$-packing of X) $\mathcal{B}$ covering B to within some infinitesimal tolerance, and by taking the 'standard part' of $\mathcal{B}$ we'll obtain a $\mathcal{S}$-tiling of X. The work comes in specifying just what this 'standard part' is, and verifying that it *is* a $\mathcal{S}$-tiling of X.

A5.5 Note The group action of $Sim\, X$ on $\mathfrak{B}X$ is topological.

Proof:

In other words we must show that the evaluation map $Sim\, X \times \mathfrak{B}X \rightarrow \mathfrak{B}X$ is continuous. Suppose $g \in \mu f$ and $B \in \mu A$ (recall that the latter means $\underline{A}^{\mu} \subseteq B \subseteq A^{\mu}$). We need to show $g B \in \mu f A$, i.e. $\underline{fA}^{\mu} \subseteq g B \subseteq (fA)^{\mu}$. As $\underline{A}^{\mu} \subseteq B$ then (using A9.1 for the third equality) $\underline{fA}^{\mu} = (f\underline{A})^{\mu} = f(\underline{A}^{\mu}) = g(\underline{A}^{\mu}) \subseteq g B$. And since $B \subseteq A^{\mu}$, $g B \subseteq g(A^{\mu}) = f(A^{\mu}) = (fA)^{\mu}$. ○

Let $\mathcal{S}^{\sim}$ denote the set of copies of elements of $\mathcal{S}$. For use in the next result, note that since any $g \in G^*$ has finite noninfinitesimal scale factor 1, $g \in G^*$ either maps $bd\, X^*$ entirely outside $bd\, X^*$ or leaves $bd\, X^*$ invariant, in which case $g \in ns\, G^*$ (using A9.2 and closure of G in the group of similitudes). Let $\mathcal{K}X$ have the Vietoris topology (given by the Hausdorff metric h recall).

A5.6 Proposition The body topology on $\mathcal{S}^{\sim}$ coincides with the Vietoris topology, and $\mathcal{S}^{\sim}$ is closed in $\mathcal{K}X$.

Proof:

We show that if $K \in \mathscr{K}X$ and $B \in (\mathscr{T}^{\sim})^*$ with $B \in \mu_{\text{vietoris}} K$, then $K \in \mathscr{T}^{\sim}$ (which shows $\mathscr{T}^{\sim}$ is closed in $\mathscr{K}X$) and $B \in \mu_{\text{body}} K$ (which shows (taking $K \in \mathscr{T}^{\sim}$ in the first place) that on $\mathscr{T}^{\sim}$ the Vietoris topology is a refinement of the body topology hence (as the latter's a refinement of the former) they're equal).

As $B \in (\mathscr{T}^{\sim})^*$ there's $S \in \mathscr{T}^*$ and $g \in G^*$ with $gS = B$. As $\mathscr{T}$ is compact in the body topology, let $T \in \mathscr{T}$ with $S \in \mu_{\text{body}} T$. Since g maps some bounded point (e.g. any element of S) to a bounded point, by the remarks preceding this proposition g is near some $f \in G$. By A5.5 we then have $B = gS \in \mu_{\text{body}} fT$, hence also $B \in \mu_{\text{vietoris}} fT$ so as $\mathscr{K}X$ is Hausdorff, $K = fT \in \mathscr{T}^{\sim}$, and we also have $B \in \mu_{\text{body}} K$. o

<u>**A5.7**</u> <u>**Note**</u> (1) There exists $\delta > 0$ such that every element of $\mathscr{T}$ expands a δ-ball.
(2) $\{\text{diam } T \mid T \in \mathscr{T}\}$ is bounded above in $(0,\infty)$.

Proof:
(1) For $A \in \mathfrak{B}X$, using compactness of A there's a largest $\delta > 0$ such that A expands a δ-ball. The map $\mathfrak{B}X \to (0,\infty)$ sending each A to its largest such δ is easily shown to be continuous, hence it achieves a minimum on compact $\mathscr{T}$.
(2) Use compactness of $\mathscr{T}$ and continuity of $\text{diam} : \mathfrak{B}X \to (0,\infty)$. Alternatively note that $\mathscr{T}$ is a compact subset of $\mathscr{K}X$, hence bounded, hence $\bigcup \mathscr{T}$ is bounded in X. o

<u>**A5.8**</u> <u>**Proposition**</u> Every $\mathscr{T}$-packing of X is closed in $\mathscr{K}X$.

Proof:
Let $\mathscr{A}$ be a $\mathscr{T}$-packing of X. Then where δ is as stated in (1) of the last note, it follows that for any distinct $A_1, A_2 \in \mathscr{A}$, $h(A_1, A_2) \geq \delta$ (because where $f \in G$ and $T \in \mathscr{T}$ with $fT = A_1$, and $[x]_\delta \subseteq T$, then $[fx]_\delta = f[x]_\delta \subseteq A_1$, so as $[fx]_\delta$ and A_2 are nonoverlapping, $d(fx, A_2) \geq \delta$). Hence $\mathscr{A}$ is closed. o

Every $\mathscr{T}$-packing of X thus belongs to $\mathscr{C}\mathscr{K}X$, which we now make into a compact Hausdorff space by giving it the S-compact topology ; so for $\mathscr{B} \in \mathscr{C}\mathscr{K}X^*$, ${}^\circ\mathscr{B} = st\mathscr{B} = \{{}^\circ B \mid B \in \mathscr{B} \text{ with } B \in ns\mathscr{K}X^*\} = \{st\, B \mid B \in \mathscr{B} \text{ with } B \subseteq ns\, X^*\} = \{st\, B \mid B \in \mathscr{B} \text{ with } B \subseteq bd\, X^*\}$.

<u>**A5.9**</u> <u>**Proposition**</u> The set of $\mathscr{T}$-packings of X is closed (i.e. compact) in $\mathscr{C}\mathscr{K}X$.

Proof:
Let $\mathscr{B}$ be a $*$-($\mathscr{T}$-packing of X); we must show that $st\mathscr{B}$ is a $\mathscr{T}$-packing of X. As $\mathscr{B} \subseteq (\mathscr{T}^{\sim})^*$, $st\mathscr{B} \subseteq st(\mathscr{T}^{\sim})^* = \mathscr{T}^{\sim}$ as $\mathscr{T}^{\sim}$ is closed. It remains to show that distinct members A_1 and A_2 of $st\mathscr{B}$ are nonoverlapping. Let $B_i \in \mathscr{B}$ with $B_i \in \mu A_i$ (with respect to the Vietoris topology, hence (by A5.6) with respect to the body topology). Then $\underline{A_i}^{\mu} \subseteq \underline{B_i}$, so as $\underline{B_1} \cap \underline{B_2} = \emptyset$ (B_1 and B_2 are nonoverlapping, noting $B_1 \neq B_2$ as $\mathscr{K}X$ is Hausdorff) we have (as $\underline{A_i} \subseteq \underline{A_i}^{\mu}$) $\underline{A_1} \cap \underline{A_2} = \emptyset$ as required. o

A5.10 **Lemma** For any $*$-($\mathcal{T}$-packing of X) $\mathcal{B}$, $\bigcup st\mathcal{B} = st \bigcup \mathcal{B}$.

Proof:

For any $B \in \mathcal{B}$ with $st\, B \neq \emptyset$, $B \subseteq bd\, X^*$ (noting that B has *finite* diameter, by A5.7(2) and the fact that the elements of G are isometries). Thus we have $\bigcup\{st\, B \mid B \in \mathcal{B} \text{ with } B \subseteq bd\, X^*\} = \bigcup\{st\, B \mid B \in \mathcal{B}\}$, i.e. $\bigcup st\mathcal{B} = st \bigcup \mathcal{B}$. o

A5.11 **Proposition** For any $*$-($\mathcal{T}$ packing of X) $\mathcal{B}$,

$$st\mathcal{B} \text{ is a } \mathcal{T}\text{-tiling of } X \Leftrightarrow st \bigcup \mathcal{B} = X,$$

Proof:

We know that $st\mathcal{B}$ is a $\mathcal{T}$-packing of X by A5.9. Thus $st\mathcal{B}$ is a $\mathcal{T}$-tiling of X iff $\bigcup st\mathcal{B} = X$, i.e. (by A5.10) $st \bigcup \mathcal{B} = X$. o

As a corollary we can note that the set of $\mathcal{T}$-tilings of X is closed (i.e. compact) in $\mathcal{CK}X$, the standard part of every $*$-($\mathcal{T}$-tiling of X) $\mathcal{B}$ being a $\mathcal{T}$-tiling of X as $st \bigcup \mathcal{B} = st\, X^* = X$. However, we now come to the main result;

A5.12 **Tiling Existence Theorem**

If every bounded subset of X can be covered by $\mathcal{T}$-packings of X to within arbitrarily small tolerances, there exists a $\mathcal{T}$-tiling of X.

Proof:

Taking $*$-bounded B expanding $bd\, X^*$, and infinitesimal $\epsilon > 0$, by transfer there's a $*$-($\mathcal{T}$-packing of X) $\mathcal{B}$ with $B \subseteq [\bigcup \mathcal{B}]_\epsilon$, hence $st \bigcup \mathcal{B} = X$, so by A5.11 $st\mathcal{B}$ is a $\mathcal{T}$-tiling of X. o

A5.13 **Corollary** If every bounded subset of X can be covered by a $\mathcal{T}$-packing of X, there exists a $\mathcal{T}$-tiling of X. o

Appendix 6 : The Body Topology

Throughout let X be a regular Hausdorff space, $\mathfrak{B}X$ be the set of bodies in X, and $\mathfrak{Bi}X$ be the set of body-interiors. As noted in 'Preliminaries', the bodies correspond bijectively with the body-interiors via the mutually inverse maps $A \mapsto \underline{A}$ and $D \mapsto \overline{D}$. In this appendix we consider a topology on $\mathfrak{B}X$ (equivalently one on $\mathfrak{Bi}X$, via the above correspondence) which takes account of interiors of sets in a way the Vietoris topology does not.

Consider a body A. A is the disjoint union of its interior $\underline{A}$ and its boundary $\partial A = A - \underline{A}$, hence A^{μ} is the disjoint union of $\underline{A}^{\mu}$ and $(\partial A)^{\mu}$, which are each partitioned into monads of course. Note that $\underline{A}^{\mu} = \bigcup\{\mu x \mid x \in X \text{ and } \mu x \subseteq A^*\}$ and $A^{\mu} = \bigcup\{\mu x \mid x \in X \text{ and } \mu x \text{ intersects } A^*\}$ (since μx intersects A^* iff $x \in \overline{A} = A$), and A^* is sandwiched in between these two sets; $\underline{A}^{\mu} \subseteq A^* \subseteq A^{\mu} = \underline{A}^{\mu} \cup (\partial A)^{\mu}$. This is all illustrated on the right. The natural question arises of whether there's a topology on $\mathfrak{B}X$ in which $B \in \mu A \Leftrightarrow \underline{A}^{\mu} \subseteq B \subseteq A^{\mu}$, so that B is sandwiched between $\underline{A}^{\mu}$ and A^{μ} just like A^*,

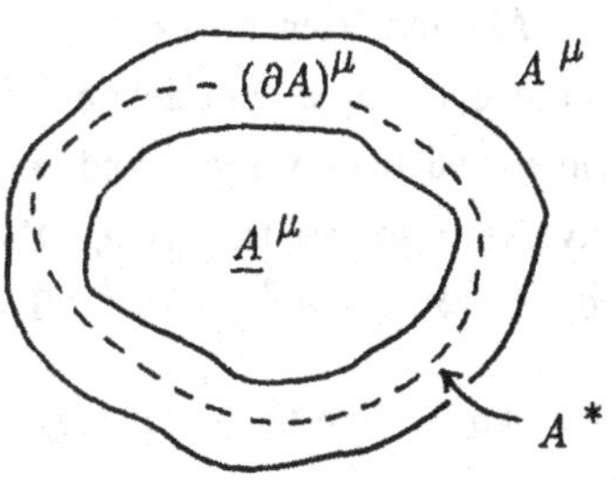

so B is substantially the same as A^*, differing only within the 'infinitesimally thin' boundary region $(\partial A)^{\mu}$. The 'body topology' will provide an affirmative answer. We point out first that in the *Vietoris* topology, $B \in \mu A$ does *not* imply that $\underline{A}^{\mu} \subseteq B$; for example, in the case $X = \mathbb{R}$ we could have $A = [-1,1]$ and $B = A^* - [i_1, i_2]$ for any infinitesimals $i_1 < i_2$. Further examples in which B is thoroughly *riddled* with infinitesimal holes within the region $\underline{A}^{\mu}$ are similarly easy to produce.

The <u>lower body topology on $\mathfrak{B}X$</u> is the topology generated by the sets of the form $[K] = \{A \in \mathfrak{B}X \mid K \subseteq \underline{A}\}$ where K is compact. The <u>upper body topology on $\mathfrak{B}X$</u> is the topology generated by the sets of the form $\text{sub } V = \{A \in \mathfrak{B}X \mid A \subseteq V\}$ where V is open. The <u>body topology on $\mathfrak{B}X$</u> is the conjunction of these, and it's easily seen that the sets of the form $[K, V] = \{A \in \mathfrak{B}X \mid K \subseteq \underline{A} \text{ and } A \subseteq V\}$ where K is compact and V is open, form a basis for the topology (noting in particular that $[K_1, V_1] \cap [K_2, V_2] = [K_1 \cup K_2, V_1 \cap V_2]$), a collection of 'basic sandwiches' if you like.

A6.1 Proposition

(1) The monads of the lower body topology on $\mathfrak{B}X$ are given by

$$\mu A = \{B \in \mathfrak{B}X^* \mid \underline{A}^{\mu} \subseteq B\}.$$

(2) The monads of the upper body topology on $\mathfrak{B}X$ are given by

$$\mu A = \{B \in \mathfrak{B}X^* \mid B \subseteq A^{\mu}\}.$$

(3) The monads of the body topology on $\mathfrak{B}X$ are given by

$$\mu A = \{B \in \mathfrak{B}X^* \mid \underline{A}^{\mu} \subseteq B \subseteq A^{\mu}\}.$$

Proof:

(1) Let $A \in \mathfrak{B}X$ and $B \in \mathfrak{B}X^*$. Suppose $B \in \mu A$. Then $\forall x \in \underline{A}$, there's a compact neighbourhood K of x with $K \subseteq \underline{A}$, so $A \in [K]$ hence $B \in [K]^*$ giving $K^* \subseteq \underline{B}$ and in particular $\mu x \subseteq B$. Thus $\underline{A}^{\mu} \subseteq B$. Conversely, suppose $\underline{A}^{\mu} \subseteq B$, which moreover implies $\underline{A}^{\mu} \subseteq \underline{B}$ (since for internal $B \subseteq X^*$, $\mu x \subseteq B \Rightarrow \mu x \subseteq \underline{B}$). Then for any K with $A \in [K]$ we have $K \subseteq \underline{A}$ so $K^* \subseteq K^{\mu} \subseteq \underline{A}^{\mu} \subseteq \underline{B}$ giving $B \in [K]^*$. So $B \in \mu A$.

(2) This just follows from 1.1(2) of Chapter 1 since the topology is the restriction to $\mathfrak{B}X$ of the sub-open topology on $\mathscr{C}X$, and as A is compact $\mu_X A = A^{\mu}$.

(3) By (1) and (2), taking the intersection of the monads. ο

Assume from now on that $\mathfrak{B}X$ has the body topology. Note that if $B \in \mu A$ then (where Δ denotes symmetric difference) $B \,\Delta\, A^* \subseteq (\partial A)^{\mu}$ since both B *and* A^* are sandwiched between $\underline{A}^{\mu}$ and A^{μ}, whose difference is $A^{\mu} - \underline{A}^{\mu} = (\partial A)^{\mu}$. In fact the converse also holds, giving the alternative characterization of the monads that $B \in \mu A \Leftrightarrow B \,\Delta\, A^* \subseteq (\partial A)^{\mu}$. The body topology is Hausdorff since . . .

A6.2 Note The body topology on $\mathfrak{B}X$ is a refinement of the Vietoris topology.

Proof:

Suppose B is near A in the body topology. We must show it's near A in the Vietoris topology, i.e. $A \subseteq st\, B$ and $B \subseteq A^{\mu}$. We already have $B \subseteq A^{\mu}$ of course, and secondly, as $\underline{A}^{\mu} \subseteq B$ we have $\underline{A} \subseteq st\, B$ hence (as A is regular and $st\, B$ is closed) $A \subseteq st\, B$. ο

$\mathfrak{B}X$ should not be expected to be locally compact; it can be shown for example that $\mathfrak{B}\,\mathbb{R}^n$ is *nowhere* locally compact. However, on the set of *convex* bodies of $\mathbb{R}^n$ the topology agrees with the Vietoris topology and is locally compact; see A8.2.

Under the bijective correspondence between $\mathfrak{B}X$ and $\mathfrak{B}\mathrm{i}X$, the body topology on the first induces a homeomorphic topology on the second (likewise with the lower and upper body topologies), which we'll naturally call the **body-interior topology on $\mathfrak{B}\mathrm{i}X$** (likewise the **lower body-interior topology on $\mathfrak{B}\mathrm{i}X$** and the **upper body-interior topology on $\mathfrak{B}\mathrm{i}X$**). The monads can be expressed as follows ;

A6.3 Proposition

(1) The monads of the lower body-interior topology on $\mathfrak{B}\mathrm{i}X$ are given by

$$\mu D = \{E \in \mathfrak{B}\mathrm{i}X^* \mid D^{\mu} \subseteq E\}.$$

(2) The monads of the upper body-interior topology on $\mathfrak{B}\mathrm{i}X$ are given by

$$\mu D = \{E \in \mathfrak{B}\mathrm{i}X^* \mid E \subseteq \overline{D}^{\mu}\}.$$

(3) The monads of the body-interior topology on $\mathfrak{B}\mathrm{i}X$ are given by

$$\mu D = \{E \in \mathfrak{B}\mathrm{i}X^* \mid D^{\mu} \subseteq E \subseteq \overline{D}^{\mu}\}.$$

Proof:
(1) We must show that for $D \in \mathfrak{B}\mathrm{i}X$ and $E \in \mathfrak{B}\mathrm{i}X^*$, $D^{\mu} \subseteq \overline{E} \Leftrightarrow D^{\mu} \subseteq E$ (as the left hand side expresses that $\overline{D} \in \mu\, \overline{E}$ in the lower body topology on $\mathfrak{B}X$). This is done as follows ; $D^{\mu} \subseteq \overline{E} \Leftrightarrow D^{\mu} \subseteq \underline{(\overline{E})} \Leftrightarrow D^{\mu} \subseteq E$ (as $\underline{(\overline{E})} = E$).
(2) We must show that $\overline{E} \subseteq \overline{D}^{\mu} \Leftrightarrow E \subseteq \overline{D}^{\mu}$. Suppose $E \subseteq \overline{D}^{\mu}$, i.e. (as $\overline{D}$ is compact, so $\overline{D}^{\mu} = \mu_X \overline{D}$) $E \subseteq \mu_X \overline{D}$. As X is regular and $\overline{D}$ is compact, $\overline{D}$ has a neighbourhood basis of *closed* sets, and it follows easily that $\overline{E} \subseteq \mu_X \overline{D}$, i.e. $\overline{E} \subseteq \overline{D}^{\mu}$.
(3) By (1) and (2), taking the intersection of the monads. o

Similar to the case of the body topology on $\mathfrak{B}X$, it can easily be shown that for $D \in \mathfrak{B}\mathrm{i}X$ and $E \in \mathfrak{B}\mathrm{i}X^*$, $E \in \mu D \Leftrightarrow E \,\Delta\, D^* \subseteq (\partial D)^*$. Also, $E \in \mu D \Rightarrow D = \mathit{subst}\, E$. And note that the sets of the form $[K,V]' = \{D \in \mathfrak{B}\mathrm{i}X \mid K \subseteq D \text{ and } \overline{D} \subseteq V\}$ where K is compact and V is open, form a basis for the body-interior topology on $\mathfrak{B}\mathrm{i}X$, namely the natural image of the basis we noted for $\mathfrak{B}X$.

Finally we give a result on the continuity of the boundary map $\partial : \mathfrak{B}X \rightarrow Cp\, X$, giving the set $Cp\, X$ of compact subsets of X the Vietoris topology (note: if $X = \mathbb{R}^n$ we can replace $Cp\, X$ by $\mathscr{K}X$ since boundaries of bodies are nonempty).

<u>A6.4</u> <u>Proposition</u> If X is locally connected, then giving $\mathfrak{B}X$ the body topology and $Cp\, X$ the Vietoris topology, the boundary map $\partial : \mathfrak{B}X \rightarrow Cp\, X$ is continuous.

Proof:
Let $A \in \mathfrak{B}X$ and $B \in \mu A$. We must show $\partial B \in \mu_{\text{vietoris}} \partial A$, i.e. (bearing in mind that $\mu_X\, \partial A = (\partial A)^{\mu}$ as ∂A is compact) $\partial B \subseteq (\partial A)^{\mu}$ and $\partial A \subseteq \mathit{st}\, \partial B$. As $\underline{A}^{\mu} \subseteq \underline{B} \subseteq B \subseteq A^{\mu}$ we have $\partial B = B - \underline{B} \subseteq A^{\mu} - \underline{A}^{\mu} = (\partial A)^{\mu}$. It remains to show that every $a \in \partial A$ belongs to $\mathit{st}\, \partial B$. As $a \in A \subseteq \mathit{st}\, B$ let $\beta \in B \cap \mu a$. As $B \in \mu A$ then $\underline{B}$ is near $\underline{A}$ in the body-interior topology, so $\underline{A} = \mathit{subst}\, \underline{B}$, i.e. $\underline{A} = \mathit{subst}\, B$, hence (as $a \notin \underline{A}$) μa is not a subset of B, so let $\gamma \in \mu a - B$. We now have $\beta, \gamma \in \mu a$, so as X is locally connected at a (i.e. a has a neighbourhood basis of connected sets) there's $*$-connected C with $\beta, \gamma \in C \subseteq \mu a$ (an elementary use of saturation). By transfer of the easy result that "for any closed $B \subseteq X$ and connected $C \subseteq X$ such that C intersects both B and $X - B$, C intersects ∂B", it follows that C intersects ∂B, so ∂B intersects μa, so $a \in \mathit{st}\, \partial B$ as required. o

Note that if we instead give $\mathfrak{B}X$ the (coarser) Vietoris topology, ∂ need not be continuous. As an example let $X = \mathbb{R}$, $A = [-1,1]$, and $B = A^* - (-\delta, \delta)$ where δ is a positive infinitesimal; then $B \in \mu_{\text{vietoris}} A$ but $\partial B \notin \mu_{\text{vietoris}} \partial A$; indeed, $\mathit{st}\, \partial B = \{-1,0,1\} = \partial A \cup \{0\} \neq \partial A$ so ∂B is not even near ∂A in the S-compact topology.

Appendix 7 : The S-Compact Topology

Throughout let X be a locally compact Hausdorff space, and $\mathscr{C}X$ denote the set of closed subsets of X. In [Nar], Narens gave a purely nonstandard description of a topology on $\mathscr{C}X$ making $\mathscr{C}X$ a compact Hausdorff space. This topology was later elaborated on by Wattenberg in [Wa], who also gave a *standard* description and dubbed the topology the 'S-compact' topology. In between the publication of these two papers the topology appeared extensively (and, so it seems, independently) in the book [Mat] of Matheron, though apparently given no name. In this appendix we'll give a brief description of the topology along with the results relevant to its application in Section 5 of Chapter 4.

Recall from Chapter 1 that the Vietoris topology on $\mathscr{C}X$ was defined as the conjunction of the 'open-intersecting' and 'closed-avoiding' topologies. The S-compact topology on $\mathscr{C}X$ on the other hand is the conjunction of the open-intersecting and compact-avoiding topologies, the latter being generated by the sets of the form **disj** $K = \{A \in \mathscr{C}X \mid A \text{ is disjoint from } K\}$. Since every compact set is closed, it follows then that the S-compact topology is a coarsening of the Vietoris topology. This can also be seen from the formulation of the monads in (3) below ;

A7.1 Proposition

(1) The monads of the open-intersecting topology on $\mathscr{C}X$ are given by

$$\mu A = \{B \in \mathscr{C}X^* \mid A \subseteq st\, B\}.$$

(2) The monads of the compact-avoiding topology on $\mathscr{C}X$ are given by

$$\mu A = \{B \in \mathscr{C}X^* \mid st\, B \subseteq A\}.$$

(3) The monads of the S-compact topology on $\mathscr{C}X$ are given by

$$\mu_s A = \{B \in \mathscr{C}X^* \mid st\, B = A\}.$$

Proof:

(1) Already proved in 1.1 of Chapter 1 ; we've listed it again for convenience.

(2) Let $B \in \mathscr{C}X^*$. First suppose $B \in \mu A$. Then $st\, B \subseteq A$ since for any $x \in st\, B$, for every compact neighbourhood K of x, B intersects K^* so $B \notin (\text{disj } K)^*$ so $A \notin \text{disj } K$, so A intersects K; so (as x has a neighbourhood basis of *compact* sets) $x \in \overline{A} = A$. Conversely suppose $st\, B \subseteq A$. Then for any compact K with $A \in \text{disj } K$ we have $B \in (\text{disj } K)^*$ (for if B *intersected* K^* then $A = st\, B$ would intersect K) ; so $B \in \mu A$.

(3) By (1) and (2), taking the intersection of the two monads. ○

As indicated in (3), we may use the notation μ_s for monads with respect to the S-compact topology. For $B \in \mathscr{C}X^*$, since B is internal then $st\, B$ is closed, and by (3) above, $B \in \mu_s\, st\, B$, i.e. ${}^\circ B = st\, B$, showing in particular that every element of $\mathscr{C}X^*$ is nearstandard, i.e. that $\mathscr{C}X$ is compact. $\mathscr{C}X$ is also Hausdorff since if $B \in \mu_s A_1 \cap \mu_s A_2$

then $A_1 = st\,B = A_2$ so $A_1 = A_2$. Also note that $\mu_s\emptyset = \{B \in \mathscr{C}X^* \mid B \subseteq rm\,X^*\}$ where $rm\,X^*$ is the set of remote (i.e. non-nearstandard) points of X^*.

For open $D \subseteq X$, the <u>D-topology on $\mathscr{C}X$</u> is the topology on $\mathscr{C}X$ induced by the map $\mathscr{C}X \to \mathscr{C}D$ (= the set of closed-in-D subsets of D) in which $A \mapsto A \cap D$, where $\mathscr{C}D$ has the S-compact topology (note that D is a locally compact Hausdorff space so the S-compact topology on $\mathscr{C}D$ *is* defined). In other words, denoting the monad of $A \in \mathscr{C}X$ with respect to the D-topology by $\mu_D A$, then for $B \in \mathscr{C}X^*$,

$$B \in \mu_D A \Leftrightarrow B \cap D^* \in \mu_s(A \cap D) \text{ in } \mathscr{C}D^*.$$

This can be expressed as follows ;

<u>A7.2</u> <u>Note</u> For open $D \subseteq X$, and $A \in \mathscr{C}X$ and $B \in \mathscr{C}X^*$,

$$\begin{aligned} B \in \mu_D A &\Leftrightarrow st(B \cap D^\mu) = A \cap D \\ &\Leftrightarrow (st\,B) \cap D = A \cap D \\ &\Leftrightarrow \forall x \in D,\ x \in st\,B \Leftrightarrow x \in A. \end{aligned}$$

Proof:
Where $st_D\,C$ denotes the standard part in D of a subset C of D^*, we have $B \in \mu_D A$ $\Leftrightarrow st_D(B \cap D^*) = A \cap D$. And as $st_D(B \cap D^*) = st(B \cap D^\mu) = (st\,B) \cap D$, the result follows. ○

Intuitively, '$B \in \mu_D A$' can be thought of as expressing that B looks like A^* within D^*; this interpretation is elaborated upon in Section 5 of Chapter 4. The S-compact topology on $\mathscr{C}X$ can now be formulated in terms D-topologies in the following way ;

<u>A7.3</u> <u>Proposition</u> For any open cover $\mathscr{D}$ of X, the S-compact topology on $\mathscr{C}X$ is the conjunction of the D-topologies on $\mathscr{C}X$ for $D \in \mathscr{D}$; in other words, for $A \in \mathscr{C}X$ and $B \in \mathscr{C}X^*$, $B \in \mu_s A \Leftrightarrow \forall D \in \mathscr{D}, B \in \mu_D A$.

Proof:
Since $\mathscr{D}$ covers X and each $D \in \mathscr{D}$ is open, $st\,B = A \Leftrightarrow \forall D \in \mathscr{D}\ (st\,B) \cap D = A \cap D$; i.e. $B \in \mu_s A \Leftrightarrow \forall D \in \mathscr{D}, B \in \mu_D A$. ○

Also note that in general, if $D_1 \subseteq D_2$ then the D_2-topology is a refinement of the D_1-topology ; this can be seen from the last formulation of '$B \in \mu_D A$' in A7.2.

Finally we give a result on limits of monotonic sequences in $\mathscr{C}X$, the proof closely following that of 4.4 of Chapter 3 ;

A7.4 **Proposition** For a sequence $\langle A_n\rangle$ in $\mathscr{C}X$,

(1) If $\langle A_n\rangle$ is decreasing, $\langle A_n\rangle \to \bigcap_n A_n$.

(2) If $\langle A_n\rangle$ is increasing, $\langle A_n\rangle \to \overline{\bigcup_n A_n}$.

Proof:

(1) Let $B = \bigcap_n A_n$. Taking any infinite m, we must show that $st\, A_m = B$. For all finite n, $A_m \subseteq A_n{}^*$ so $st\, A_m \subseteq st\,(A_n{}^*) = A_n$; so $st\, A_m \subseteq B$. Conversely, $B^* \subseteq A_m$ so $B = st\, B^* \subseteq st\, A_m$.

(2) Let $B = \overline{\bigcup_n A_n}$. Taking any infinite m, we show $st\, A_m = B$. Since $A_m \subseteq B^*$, $st\, A_m \subseteq st\, B^* = \overline{B} = B$. Conversely, for all finite n, $A_n{}^* \subseteq A_m$ so $A_n = st\,(A_n{}^*) \subseteq st\, A_m$; so $\bigcup_n A_n \subseteq st\, A_m$, so as the latter's closed, $B \subseteq st\, A_m$. ○

Appendix 8 : The Hyperspace of Convex Bodies

Throughout let $X = \mathbb{R}^n$. Let $\mathfrak{CB}X$ denote the set of convex bodies in X, namely the convex regular nonempty compact sets, and give $\mathfrak{CB}X$ the Hausdorff metric. The main result in this appendix will show that $\mathfrak{CB}X$ is locally compact and that its topology, namely the Vietoris topology, coincides with the body topology and the S-compact topology (respectively described in Appendices 6 and 7), from which we may conclude that the topology is *the* natural topology on $\mathfrak{CB}X$. As usual we make use of nonstandard analysis, which in fact finds rewarding application to the theory of convex sets (as yet unreported in the literature so far as I'm aware) due to the attractive interplay between convex subsets of X and convex subsets of X^*.

Generalising notation used in the case of $\mathbb{R}$, for $x,y \in X$, $[x,y]$ denotes the line segment with endpoints x and y, whilst $(x,y]$ denotes $[x,y] - \{x\}$.

A8.1 Lemma For $*$-convex C, $st\ C$ is a closed convex set with interior $subst\ C$.
Proof:
As C is internal, $st\ C$ is closed and $subst\ C$ is open. And $st\ C$ is convex since for all $x,y \in st\ C$ there exist $a,b \in C$ with $a \in \mu x$ and $b \in \mu y$, giving $[x,y] = st\,[a,b] \subseteq st\ C$ as $[a,b] \subseteq C$. We now show $\underline{st\ C} = subst\ C$. Since $subst\ C$ is an open subset of $st\ C$ it remains to show $\underline{st\ C} \subseteq subst\ C$. For $x \in \underline{st\ C}$ there's an open ball V on x with $V \subseteq st\ C$, so $\forall v \in V$ μv contains a point of C, so by convexity of C clearly $\mu x \subseteq C$ (indeed, $V^{\mu} \subseteq C$), so $x \in subst\ C$ as required. ○

A8.2 Proposition $\mathfrak{CB}X$ is locally compact, and its topology coincides with the body topology and the S-compact topology.
Proof:
Give the set $\mathscr{K}X$ of nonempty compact subsets of X the Hausdorff metric. The set $\mathfrak{C}X$ of *convex* elements of $\mathscr{K}X$ is closed in $\mathscr{K}X$; for if $A \in \overline{\mathfrak{C}X}$ there's $B \in \mathfrak{C}X^*$ near A, so $A = st\ B$ which is convex as B is $*$-convex. In turn $\mathfrak{CB}X$ is *open* in $\mathfrak{C}X$; for if $A \in \mathfrak{CB}X$ and $B \in \mathfrak{C}X^*$ with $B \in \mu A$, $subst\ B = \underline{st\ B} = \underline{A} \neq \emptyset$ hence $\underline{B} \neq \emptyset$ (as $(subst\ B)^{\mu} \subseteq \underline{B}$), so B is $*$-regular, hence belongs to $\mathfrak{CB}X^*$. Now local compactness is closed-hereditary and open-hereditary, hence as $\mathscr{K}X$ is locally compact, so is $\mathfrak{C}X$, and in turn so is $\mathfrak{CB}X$.

Next we show that the body topology and S-compact topology coincide with the Vietoris topology on $\mathfrak{CB}X$. Since the body topology is a refinement of the Vietoris topology which is a refinement of the S-compact topology, it remains to complete the circle and show that on $\mathfrak{CB}X$, the S-compact topology is a refinement of the body topology. So, suppose $A \in \mathfrak{CB}X$ and $B \in \mathfrak{CB}X^*$ with $B \in \mu_s A$. We must show B is near A in the body topology, i.e. $\underline{A}^{\mu} \subseteq B \subseteq A^{\mu}$. Firstly, $B \subseteq ns\, X^*$; for if this were false, say $b \in B - ns\, X^*$, then taking any $c \in B \cap ns\, X^*$ (such c exists as $st\, B = A \neq \emptyset$) we have $[b,c] \subseteq B$ so $st\,[b,c] \subseteq st\, B \subseteq A$, but $st\,[b,c]$ is unbounded, contradicting that A is bounded. We now have $B \subseteq ns\, X^*$ and $st\, B = A$, i.e. B is near A in the Vietoris topology, which in particular gives $B \subseteq A^{\mu}$. Lastly, $\underline{A}^{\mu} \subseteq B$ since $\underline{A} = \underline{st\, B} = subst\, B$ using A8.1. ○

Note that since the set of closed balls is closed in $\mathfrak{CB}X$ (quite easily proved nonstandardly), it too is locally compact. It should be noted incidentally that $\mathfrak{CB}X$ is not *boundedly* compact ; for example we can have $B \in \mathfrak{CB}X^*$ with $B \subseteq \mu x$ for some x, hence in $\mathscr{K}X^*$ $B \in \mu\{x\}$ so B is *not* near any element of $\mathfrak{CB}X$.

Let $\mathfrak{CBi}X$ denote the set of *interiors* of convex bodies, equivalently the set of convex body-interiors. $\mathfrak{CB}X$ and $\mathfrak{CBi}X$ correspond bijectively under the mutually inverse maps $A \mapsto \underline{A}$ and $D \mapsto \overline{D}$ (using the facts that interior and closure both preserve convexity), and we naturally give $\mathfrak{CBi}X$ the topology which makes it homeomorphic to $\mathfrak{CB}X$ under this correspondence ; this is namely the body-interior topology inherited from $\mathfrak{Bi}X$ (see Appendix 6). $\mathfrak{CBi}X$ is thus locally compact, as is the set of open balls (namely comprising the interiors of the *closed* balls).

<u>A8.3</u> <u>**Lemma**</u> For $*$-convex C with $subst\, C \neq \emptyset$, $st\, C$ is the closure of $subst\, C$, so $st\, C$ and $subst\, C$ are corresponding nonempty *regular* convex sets. If also $C \subseteq ns\, X^*$, $st\, C$ and $subst\, C$ are corresponding elements of $\mathfrak{CB}X$ and $\mathfrak{CBi}X$.

Proof:

Firstly we show $st\, C = \overline{subst\, C}$. $st\, C$ is a closed expansion of $subst\, C$, so it remains to show that $st\, C \subseteq \overline{subst\, C}$. Let $x \in st\, C$; say $c \in C \cap \mu x$. Taking some $y \in subst\, C$, we obtain that $(x,y]^{\mu} \subseteq C$, because for $z \in (x,y]$ there's $b \in [c,y]$ with $b \approx z$, and necessarily $b \not\approx c$, hence $b^{\approx} \subseteq C$ (as $b^{\approx}$ is the image of $\mu y \subseteq C$ under the dilation about c taking y to b), i.e. $\mu z \subseteq C$. So $(x,y] \subseteq subst\, C$, giving $x \in \overline{subst\, C}$ as required. By A8.1 we now have that $st\, C$ and $subst\, C$ are corresponding *regular* nonempty convex sets. If also $C \subseteq ns\, X^*$, then $st\, C$ is compact, hence $st\, C \in \mathfrak{CB}X$, and correspondingly $subst\, C \in \mathfrak{CBi}X$. ○

The following gives a simple formulation of the monads of $\mathfrak{CBi}X$;

<u>A8.4</u> <u>**Proposition**</u> For $D \in \mathfrak{CBi}X$, $\mu D = \{E \in \mathfrak{CBi}X^* \mid subst\, E = D\}$.

Proof:

For $E \in \mathfrak{CBi}X^*$ we must show $E \in \mu D \Leftrightarrow subst\, E = D$. Both sides give $subst\, E \neq \emptyset$, so now assume this. Hence by A8.3 $st\, E$ and $subst\, E$ are corresponding regular sets, and we can reason as follows ;

$E \in \mu D \Leftrightarrow \overline{E} \in \mu \overline{D}$	by the homeomorphism of $\mathfrak{CBi}X$ with $\mathfrak{CB}X$
$\Leftrightarrow st\,\overline{E} = \overline{D}$	since $\mathfrak{CB}X$ has the S-compact topology
$\Leftrightarrow st\,E = \overline{D}$	as $st\,\overline{E} = st\,E$
$\Leftrightarrow subst\,E = D$	by taking interiors. ○

Lastly, making use of the following result on continuity of the convex hull operation $conv : \mathscr{K}X \to \mathscr{K}X$, we show that $\mathfrak{CB}X$ is embedded in $\mathscr{K}X$ by ∂;

<u>A8.5</u> <u>Lemma</u> The convex hull operation $conv : \mathscr{K}X \to \mathscr{K}X$ is uniformly continuous with respect to the Hausdorff metric.

Proof:

For $A,B \in (\mathscr{K}X)^*$ with $A \approx B$, we must show that $conv\,A \approx conv\,B$. By transfer of a formula for convex hulls, $conv\,A = \{\lambda_1 a_1 + \cdots + \lambda_{n+1} a_{n+1} \mid \text{all } a_i \in A \text{ and } \lambda_i \in [0,1]^* \text{ with } \lambda_1 + \cdots + \lambda_{n+1} = 1\}$. For any $c = \lambda_1 a_1 + \cdots + \lambda_{n+1} a_{n+1}$ in $conv\,A$, for each i there's $b_i \in B$ with $b_i \approx a_i$, and for such b_i we then have $c \approx \lambda_1 b_1 + \cdots + \lambda_{n+1} b_{n+1} \in conv\,B$. Likewise every element of $conv\,B$ is infinitesimally distant from an element of $conv\,A$. Thus $conv\,A \approx conv\,B$. ○

<u>A8.6</u> <u>Proposition</u> Giving $\mathscr{K}X$ the Vietoris topology, the boundary map $\partial : \mathfrak{CB}X \to \mathscr{K}X$ is an embedding.

Proof:

By A6.4 the map ∂ above is continuous. Now ∂ is also injective since for $A \in \mathfrak{CB}X$, $A = conv\,\partial A$. And for $A \in \mathfrak{CB}X$ and $B \in (\mathfrak{CB}X)^*$, if $\partial B \in \mu\,\partial A$ then by A8.5 $conv\,\partial B \in \mu\,conv\,\partial A$, i.e. $B \in \mu A$. ○

Appendix 9 : Similitudes

In this appendix we provide a summary of the main facts about 'similitudes' of $X = \mathbb{R}^n$ used in the monograph, including a few nonstandard results.

A **similtude** of X is a bijection $f: X \to X$ for which there exists $r \in (0,\infty)$ such that $\forall x,y \in X \;\; d(fx,fy) = r\, d(x,y)$. This r is unique and is called the **scale factor** of f (it is the 'Lipschitz ratio' r_f of f too of course). The similitudes form a group ***Sim X*** of homeomorphisms of X, and the scale factor map $r: \mathit{Sim}\, X \to (0,\infty)$ is a group homomorphism. Under the compact-open topology, assumed throughout, *Sim X* forms a *topological* group (with the scale factor map becoming continuous). Continuity of composition holds simply from that in $C(X,X)$, whilst continuity of the inverse map $f \mapsto f^{-1}$ can easily be proved with the aid of the following nonstandard result A9.1 for example. First we define that a bijection $g: X^* \to X^*$ is **strongly microcontinuous** if $\forall \alpha,\beta \in X^* \;\; \alpha \approx \beta \Leftrightarrow g\alpha \approx g\beta$, equivalently if g and g^{-1} are microcontinuous, equivalently if $\forall \alpha \in X^* \;\; g(\alpha^{\approx}) = (g\alpha)^{\approx}$. Note then that a $*$-similitude g is strongly microcontinuous iff r_g is finite and noninfinitesimal (which in particular holds if g is nearstandard in $(\mathit{Sim}\, X)^*$, by continuity of the scale factor map).

A9.1 Note For $g \in (\mathit{Sim}\, X)^*$ and $f \in \mathit{Sim}\, X$, $g \in \mu f \Leftrightarrow \forall x \in X \;\; g\,\mu x = f\,\mu x$.

Proof:

$\Rightarrow$: Let $x \in X$. Since g is strongly microcontinuous (being nearstandard) and $g\,x \approx f\,x$, $\forall \alpha \in X^*$, $\alpha \in \mu x \Leftrightarrow \alpha \approx x \Leftrightarrow g\,\alpha \approx g\,x \Leftrightarrow g\,\alpha \approx f\,x \Leftrightarrow g\,\alpha \in \mu\, f\,x$. So $g\,\mu x = \mu\, f\, x$, which is $f\,\mu x$.

$\Leftarrow$: Then $\forall x \in X \;\; g\,\mu x \subseteq \mu\, f\,x$; i.e. $g \in \mu f$. ○

A9.2 Proposition For $g \in (\mathit{Sim}\, X)^*$, the following are equivalent ;

(1) g is nearstandard in $(\mathit{Sim}\, X)^*$.

(2) $g\, bd\, X^* = bd\, X^*$.

(3) r_g is finite and noninfinitesimal, and g maps some bounded point to a bounded point.

Proof:

(1) $\Rightarrow$ (2) : $g\, bd\, X^* = g(X^{\mu}) = \bigcup_x g\,\mu x = \bigcup_x f\,\mu x = \bigcup_x \mu\, f\,x = X^{\mu} = bd\, X^*$.

(2) $\Rightarrow$ (3) : Since $g\, bd\, X^* \subseteq bd\, X^*$, r_g is finite (else g could only map at most one element of X to a bounded point). And since $bd\, X^* \subseteq g\, bd\, X^*$, r_g is noninfinitesimal.

(3) $\Rightarrow$ (1) : Since r_g is finite and g maps a bounded point to a bounded point, we have $g\, bd\, X^* \subseteq bd\, X^*$. By the same reasoning applied to g^{-1} (noting that $r_{(g^{-1})} = (r_g)^{-1}$ which is thus finite and noninfinitesimal) we have $g^{-1} bd\, X^* \subseteq bd\, X^*$, i.e. $bd\, X^* \subseteq g\, bd\, X^*$. So $g\, bd\, X^* = bd\, X^*$. Since g is also stongly microcontinuous, g maps

monads to monads, and letting f be the unique map $X \to X$ such that $\forall x \in X$ $g\,\mu x = \mu\, f x$, it is easy to show that f is a similitude with scaling ratio ${}^{\circ}r_g$. By the last note we then have $g \in \mu f$. ○

The definition of a similitude given above was purely metric, but similitudes can be resolved into certain 'geometric' transformations as follows. Firstly, an **orthogonal map** of X is a linear isometry of X, and these are in fact the only isometries of X leaving 0 fixed. For $p \in X$ and $r \in (0,\infty)$, the **dilation about p by factor r** is the map f defined by $f x = r(x-p) + p$ (note: if $r < 1$ we may refer to f as the *contraction* about p by factor r), and a map of this form is called a **dilation**. It can be shown that any similitude f can be expressed uniquely in the form 'orthogonal map followed by dilation about 0 followed by translation'. We say f is **direct** if the orthogonal map involved is a **rotation**, i.e. has determinant 1. Letting ***Orthog X*** be the group of orthogonal maps, ***$Dil_0 X$*** the group of dilations about 0, and ***Trans X*** be the group of translations of X, all these being topological subgroups of *Sim X*, the map *Trans X* × *Dil_0X* × *Orthog X* → *Sim X* in which $(f, g, h) \mapsto f \circ g \circ h$ is a homeomorphism. And in the natural way, *Trans X* ≅ X and *Dil_0X* ≅ $(0,\infty)$ of course, so *Sim X* ≅ $X \times (0,\infty) \times$ *Orthog X*. It follows that *Sim X* is locally compact since each of the three factors on the right are (*Orthog X* is compact moreover). The group of *direct* similitudes (namely generated by the translations, dilations, and rotations) is closed in *Sim X*, hence likewise locally compact.

References

[AFHL] Albeverio, Fenstad, Høegh-Krohn, & Lindstrøm, *Nonstandard Methods in Stochastic Analysis and Mathematical Physics*, Academic Press, 1986.

[Cu] N. J. Cutland, *Nonstandard Measure Theory and its Applications*, Bull. London Math. Soc., **15** (1983), 529-589.

[Da] M. Davis, *Applied Nonstandard Analysis*, John Wiley & Sons, 1977.

[DK] R. Devaney & L. Keen (Eds.), *Chaos and Fractals : the Mathematics Behind the Computer Graphics*, Proceedings of Symposia in Applied Mathematics, Vol 39, A.M.S.

[Du] J. Dugundji, *Topology*, Allyn and Bacon Inc., 1966.

[Ed] M. Edelstein, *On Fixed and Periodic Points under Contractive Mappings*, J. London. Math. Soc. **37** (1962) 74-79.

[Fa] K. J. Falconer, *The Geometry of Fractal Sets*, C.U.P. 1985.

[GS] B. Grünbaum & G. Shephard, *Tilings and Patterns ; an Introduction* , W. H. Freeman, New York, 1989.

[Ha] M. Hata, *On the Structure of Self-Similar Sets*, Japan J. Appl. Math., **2** (1985), 381-414.

[Haus] F. Hausdorff, *Mengenlehre*, Dover, New York, 1944.

[HL] A. E. Hurd & P. A. Loeb, *An Introduction to Nonstandard Real Analysis*, Academic Press, 1985.

[Hu] J. E. Hutchinson, *Fractals and Self-Similarity*, Indiana Univ. Math. J., **30** (1981), 713-747.

[HW] W. Hurewicz & H. Wallman, *Dimension Theory*, Princeton Univ. Press, 1941

[Ju] I. Juhász, *Non-standard Notes on the Hyperspace*, pp. 171-177 of *Contributions to Nonstandard Analysis*, W. A. J. Luxemburg & A. Robinson (Eds.), North-Holland, 1972.

[Ke] H. J. Keisler, *Foundations of Infinitesimal Calculus*, Prindle, Weber & Schmidt, Massachusetts, 1976.

[Ku] K. Kuratowski, *Topology* Vols 1 & 2, Academic Press, New York, 1966.

[Li,1] T. Lindstrøm, *A Nonstandard Approach to Iterated Function Systems*, preprint.

[Li,2] T. Lindstrøm, *Brownian Motion on Nested Fractals*, Memoirs of the A.M.S., Vol 83, No. 420 (1990).

[Li,3] T. Lindstrøm, *An Invitation to Nonstandard Analysis*, in *Nonstandard Analysis and its Applications*, N. J. Cutland (Ed.), C.U.P. 1988.

[Man] B. B. Mandelbrot, *The Fractal Geometry of Nature*, W. H. Freeman, New York, 1983.

[Mar] G. Martin, *Transformation Geometry*, Springer, 1982.

[Mat] G. Matheron, *Random Sets and Integral Geometry*, John Wiley & Sons, 1975.

[Mi] E. Michael, *Topologies on Spaces of Subsets*, Trans. Amer. Math. Soc., **71** (1951), 152-182.

[Mu] M. G. Murdeshwar, *General Topology*, Wiley Eastern Limited, 1983.

[Nad] S. Nadler, *Sequences of Contractions and Fixed Points*, Pacific. J. Math., **27** (1968) No.3, 579-585.

[Nar] L. Narens, *Topologies of Closed Subsets*, Trans. Amer. Math. Soc., **174** (1972) 55-76.

[Ro] A. Robinson, *Non-standard Analysis*, North-Holland, 1966.

[SL] K. D. Stroyan & W. A. J. Luxemburg, *Introduction to the Theory of Infinitesimals*, Academic Press, 1976.

[Vi] L. Vietoris, *Bereiche Zweiter Ordnung*, Monatschefte für Mathematik und Physik, **32** (1922), 258-280.

[Wa] F. Wattenberg, *Topologies on the Set of Closed Subsets*, Pacific. J. Math., **68** (1977) No.2, 537-551.

[Wic] K. R. Wicks, *Spiral-Based Self-Similar Sets*, Mathematics Research Reports, University of Hull, Vol 3 (1990) No. 1. Also to be published in *Spiral Symmetry* (Eds. I. Hargittai & C. A. Pickover), World Scientific Ltd.

[Wil] R. F. Williams, *Composition of Contractions*, Bol. Soc. Brasil Mat. **2** (1971) 55-59.

Notation Index

Preliminaries

Chapter 1 : Nonstandard Development of the Vietoris Topology

Chapter 2 : Nonstandard Development of the Hausdorff Metric

Chapter 3 : Hutchinson's Invariant Sets

Chapter 4 : Views and Fractal Notions

Appendices

Term Index